NASA STI Program . . . in Profile

Since its founding, NASA has been dedicated to the advancement of aeronautics and space science. The NASA Scientific and Technical Information (STI) program plays a key part in helping NASA maintain this important role.

The NASA STI Program operates under the auspices of the Agency Chief Information Officer. It collects, organizes, provides for archiving, and disseminates NASA's STI. The NASA STI program provides access to the NASA Aeronautics and Space Database and its public interface, the NASA Technical Reports Server, thus providing one of the largest collections of aeronautical and space science STI in the world. Results are published in both non-NASA channels and by NASA in the NASA STI Report Series, which includes the following report types:

- TECHNICAL PUBLICATION. Reports of completed research or a major significant phase of research that present the results of NASA programs and include extensive data or theoretical analysis. Includes compilations of significant scientific and technical data and information deemed to be of continuing reference value. NASA counterpart of peer-reviewed formal professional papers but has less stringent limitations on manuscript length and extent of graphic presentations.

- TECHNICAL MEMORANDUM. Scientific and technical findings that are preliminary or of specialized interest, e.g., quick release reports, working papers, and bibliographies that contain minimal annotation. Does not contain extensive analysis.

- CONTRACTOR REPORT. Scientific and technical findings by NASA-sponsored contractors and grantees.

- CONFERENCE PUBLICATION. Collected papers from scientific and technical conferences, symposia, seminars, or other meetings sponsored or cosponsored by NASA.

- SPECIAL PUBLICATION. Scientific, technical, or historical information from NASA programs, projects, and missions, often concerned with subjects having substantial public interest.

- TECHNICAL TRANSLATION. English-language translations of foreign scientific and technical material pertinent to NASA's mission.

Specialized services also include creating custom thesauri, building customized databases, organizing and publishing research results.

For more information about the NASA STI program, see the following:

- Access the NASA STI program home page at *http://www.sti.nasa.gov*

- E-mail your question via the Internet to *help@sti.nasa.gov*

- Fax your question to the NASA STI Help Desk at 443–757–5803

- Telephone the NASA STI Help Desk at 443–757–5802

- Write to:
  NASA Center for AeroSpace Information (CASI)
  7115 Standard Drive
  Hanover, MD 21076–1320

NASA/TM—2011-216972

CD–2007–19

COMPASS Final Report: Saturn Moons Orbiter Using Radioisotope Electric Propulsion (REP): Flagship Class Mission

Steven R. Oleson and Melissa L. McGuire
Glenn Research Center, Cleveland, Ohio

National Aeronautics and
Space Administration

Glenn Research Center
Cleveland, Ohio 44135

February 2011

Acknowledgments

The COMPASS team wishes to thank the support of the Engineering Directorate, and specifically branch Chief Maria Babula and mission analysts of the Mission Design and Analysis Branch for their unending support and high caliber of engineer support. The team would also like to thank the In Space Propulsion Office for always providing challenging design study ideas.

Level of Review: This material has been technically reviewed by technical management.

Available from

NASA Center for Aerospace Information	National Technical Information Service
7115 Standard Drive	5301 Shawnee Road
Hanover, MD 21076–1320	Alexandria, VA 22312

Contents

COMPASS Final Report: Saturn Moons Orbiter Using Radioisotope Electric Propulsion (REP): Flagship Class Mission

Steven R. Oleson and Melissa L. McGuire
National Aeronautics and Space Administration
Glenn Research Center
Cleveland, Ohio 44135

1.0   Executive Summary

The COllaborative Modeling and Parametric Assessment of Space Systems (COMPASS) team was approached by the NASA Glenn Research Center (GRC) In-Space Project to perform a design session to develop Radioisotope Electric Propulsion (REP) Spacecraft Conceptual Designs (with cost, risk, and reliability) for missions of three different classes: New Frontier's Class Centaur Orbiter (with Trojan flyby), Flagship, and Discovery. The designs will allow trading of current and future propulsion systems. The results will directly support technology development decisions. The results of the Flagship mission design are reported in this document (see Figure 1.1).

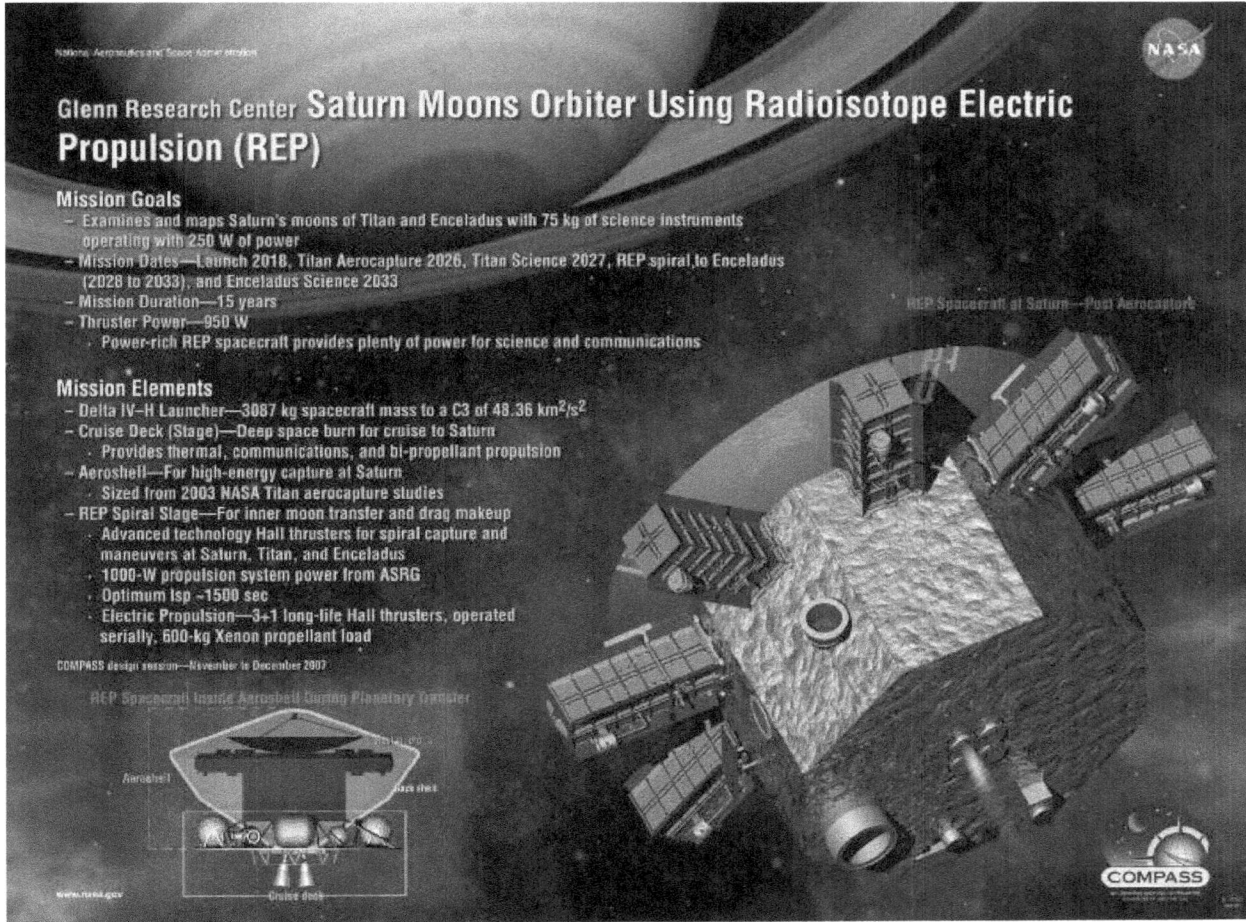

Figure 1.1.—Mission overview.

The mission chosen for this design is a science mission to the Saturn system to map Saturn and its moons Titan and Enceladus using REP to power Electric Propulsion (EP) thrusters and the S/C and science instruments. The requirement for the mission is to optimize the mass of a science orbiter delivered to the Saturn moon system. The REP S/C is launched in a Delta IV H. The S/C consists of three "stages": a chemical cruise deck is used to perform one mid-course burn, an aeroshell is used to break into Titan orbit, and the Radioisotope-powered EP thrusters are used for operation between the moons and for atmospheric drag compensation about the moon Titan. After a chemical burn using the cruise stage, the aeroshell performs the aerocapture maneuver at Titan, and once at Titan, the REP powered EP stage provides propulsion for drag makeup at Titan and a 5-yr transfer from Titan to Enceladus. Figure 1.2 shows the conceptual REP Vehicle with chemical propulsion cruise deck (stage) and aeroshell designed for this COMPASS session.

In Figure 1.2, the Cruise Deck (stage) is shown at the bottom, and the REP stage is shown inside the aeroshell. The major systems (heat shield, back shell) and stages (Cruise Deck) of the complete vehicle are labeled.

Table 1.1 summarizes the Cruise Stage/Aerocapture/REP stage S/C and mission.

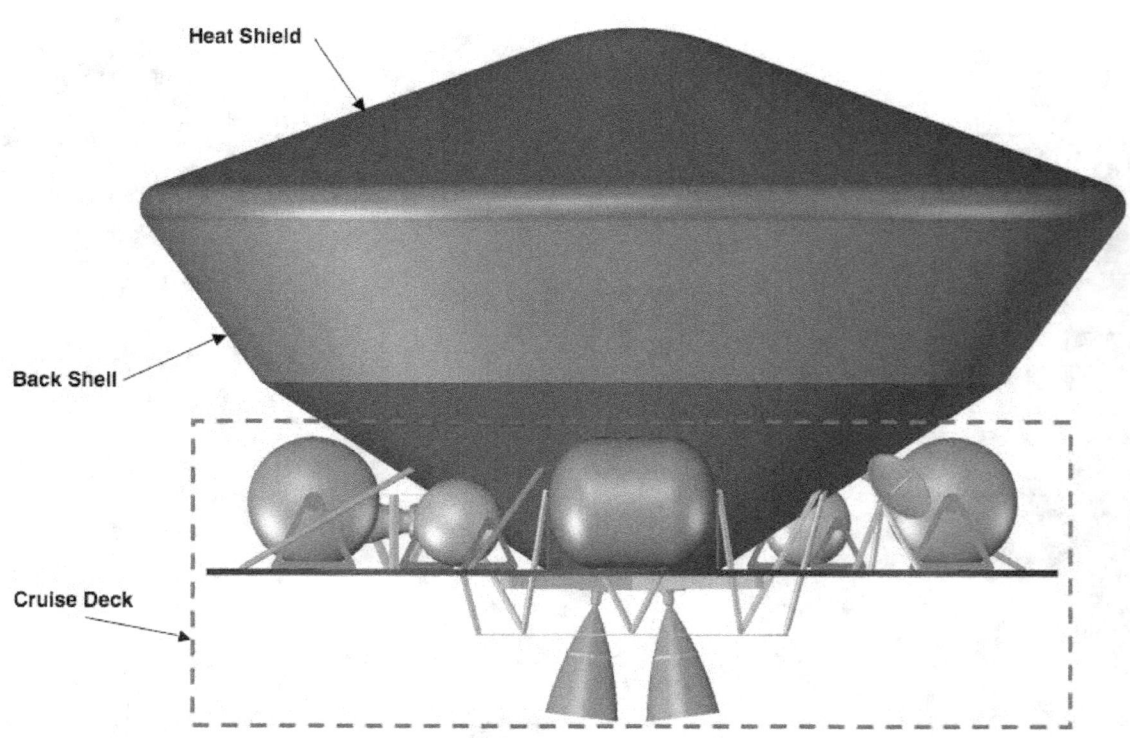

Figure 1.2.—Conceptual design REP science spacecraft (S/C) with Cruise Deck (REP stage inside aeroshell).

TABLE 1.1.—MISSION AND S/C SUMMARY

Mission	Titan Aerocapture in 8 yr, 1 yr Titan REP Drag makeup, 5 yr spiral down to Enceladus, 7 km/s,	Total mass with growth
System	30% system level growth (additional 105 kg carried at system level), American Institute for Aeronautics and Astronautics (AIAA) mass growth allowance (MGA) schedule used at subsystem level	3087 kg (wet)
Launch	Launch: Delta IV Heavy, C_3 48 km^2/s^2, 10 kg Expendable Launch Vehicle (ELV) adaptor, 49 kg available launch margin	3136 kg performance
Science	Science: 70 kg, 250 W (based on Applied Physics Laboratory (APL) Titan flagship payload)	91 kg
Power	Eight Advanced Stirling Radioisotope Generators (ASRG) with multilayer insulation (MLI), attached (loaded) in pairs, 1200 W end of life (EOL) (14-yr)	249 kg
Chemical Cruise Deck	Cruise Deck provides deep space maneuver (bipropellant), thermal and communications during cruise to Saturn	954 kg (wet)
Aeroshell	Aerocapture system sized from 2003 NASA Titan Aerocapture Studies, Aeroshell included in the REP Stage masses	315 kg
Mechanical	Octagonal Al-Li bus with Propulsion and Science Decks, capable of carrying 6 g axial and 3.5 g lateral launch loads	100 kg
Electric Propulsion	3+1 Long Life Hall Thrusters, operated serially, 600 kg Xe propellant load 900 W into thruster, 1500 s, Specific Impulse (I_{sp}), 200 kg Xe throughput each, 30,000 hr Single string power processing unit (PPU) (95%), thruster feed, thruster	187 kg (dry)
Command and Data Handling (C&DH), Communications	RAD 750, 220 W, 10 kb/s Data rate, Ka Band, 2.1 m antenna	104 kg
Attitude Determination and Control (AD&C)	Two Star Cameras, Inertial Measuring Unit (IMU), four reaction wheels, Hydrazine Reaction Control System (RCS) propulsion system	30 kg
Thermal	Hibernate S/C during interplanetary coast, Aerocapture shell consists of active radiators, post shell separation MLI and heaters	127 kg

2.0 Study Background and Assumptions

2.1 Introduction

The Executive Summary provides a framework for where this study fits into NASA science mission goals. The rest of the document addresses the issues of: mission category, propulsion type and engines, mission duration, scientific payload and its requirements, power requirements, communications requirements, launch vehicle and a specific mission targets (Titan and Enceladus). As with any study, there are many options among these mission elements. Many of them are traded for comparison in order to provide optimal scientific results, minimize cost and provide the highest probability of mission success (as defined by established goals). There is a preliminary discussion on past missions that serve as baseline designs.

2.2 Purpose

The goal of this study is to determine a preliminary S/C design for a Flagship class mission to study Saturn's moons Titan and Enceladus focusing on the application of REP technology to enable and enhance a large scale, high power science mission.

2.3 Assumptions and Approach

The following section contains the description of the mission class that was the subject of this design study. The following details are the most current available at the time of this design session.

This study will utilize the COMPASS S/C Conceptual Design team to provide complete Science Class Reference Mission Designs.

2.3.1 NASA Flagship Class Mission Definition

NASA Planetary Science Division (PSD) is developing plans for the next outer planet moon flagship mission. The moons under consideration are: Europa, Titan, Enceladus, and Ganymede.

The Definition of a Flagship Class Mission is as follows (Ref. 1):

The largest missions, known as Flagship Missions, range in cost from $800 to $1400 million or $1400 to $2800 million. These missions will be crucial in allowing us to reach and explore difficult but high-priority targets. These critically important targets could help establish the limits of habitability, not just for our solar system, but also for planetary systems in general. In particular, they potentially provide an opportunity to identify pre-biotic organic molecules or even extant life beyond Earth, should it exist, in our own solar system. The targets of flagship missions include complex missions to the surface of Venus, the lower atmosphere and surface of Titan, the surface and subsurface of Europa, the deep atmosphere of Neptune and the surface of its moon Triton, and the surface of a comet nucleus in the form of cryogenically preserved samples.

2.3.2 Design Starting Points

The design began by comparing the performance of a launch on an Atlas 551/Star 48 and a Delta IV H class ELV on a direct injection trajectory to Saturn. A simple trade was performed to start the session, between the uses of the ASRG powered electric thrusters operating to Saturn versus the use of a chemical cruise stage (total ΔV to provide ~2.5 km/s). The chemical cruise stage was chosen as simpler since it avoids operating the EP thrusters while inside the aeroshell. The following is a list of goals of the S/C mission design

(1) Transfer optimal mass from Earth to Saturn
(2) Aerocapture at Titan with an aeroshell integrated around the REP stage
(3) Perform 1 yr mission of mapping Titan surface from an orbit of between 1200 and 1400 km using the ASRG power for science and S/C instruments, and for orbit maintenance (budgets $\Delta V \sim 100$ m/s)
(4) Use REP stage to spiral out of Titan orbit.
(5) Use the REP stage to spiral down to Enceladus and
(6) Spend 1 yr at Enceladus doing science mapping.

2.4 Growth, Contingency and Margin Policy

Mass Growth: For dry mass elements in the system design, the COMPASS team uses the ANSI/AIAA R-020A-1999, "Recommended Practice for Mass Properties Control for Satellites, Missiles, and Launch Vehicles," (Ref. 2). Table 2.1 shows the Percent Mass Growth separated into a matrix specified by level of design maturity and specific subsystem.

The percent growth factors are applied to each subsystem, after which the total system growth of the vehicle level is calculated. The COMPASS team desired total growth to be 30 percent, and an additional growth is carried at the system level in order to achieve a total system growth of a 30 percent limit on the dry mass of the system. Note that for designs requiring propellant, growth in propellant is either book kept in the propellant itself or in the ΔV used to calculate the propellant necessary to fly a mission.

The COMPASS team uses the Discover Announcement of Opportunity (AO) definitions of Contingency and mass Margin.

From the Discovery AO: Definitions of Contingency and Mass Margin

Contingency (or Reserve), when added to a resource, results in the maximum expected value for that resource. Percent contingency is the value of the contingency divided by the value of the resource less the contingency.

Margin is the difference between the maximum possible value of a resource (the physical limit or the agree-to limit) and the maximum expected value for a resource. Percent margin for a resource is the available margin divided by its maximum expected value.

Power Growth: The COMPASS team uses a 30 percent power growth assumption except in cases of electric propulsion since any reduction in power can be handled by increasing trip time.

TABLE 2.1.—PERCENT MASS GROWTH ALLOWANCE

Code	Design Maturity (Basis for Mass Determination)	Percent Mass Growth Allowance									
		Electrical/Electronic Components			Structure	Thermal Control	Propulsion	Batteries	Wire Harnesses	Mechanisms	Instrumentation
		0-5 kg	5-15 kg	>15 kg							
E	Estimated (preliminary sketches)	30	20	15	18	18	18	20	50	18	50
L	Layout (or major modification of existing hardware)	25	20	15	12	12	12	15	30	12	30
P	Pre-Release Drawings (or minor modification of existing hardware)	20	15	10	8	8	8	10	25	8	25
C	Released Drawings (calculated values)	10	5	5	4	4	4	5	5	4	5
X	Existing Hardware (actual mass from another program)	3	3	3	2	2	2	3	3	2	3
A	Actual Mass (measured flight hardware)	0	0	0	0	0	0	0	0	0	0
CFE	Customer Furnished Equipment	0	0	0	0	0	0	0	0	0	0

Following the conventions of CD–2007–16 (Ref. 3), the APL assumptions on science payload margins are applied to this science payload.

Science payload mass and power margins assumptions are:

- 30 percent margin should be *added* for mass estimates (mass estimates do not include any additional shielding for the REP mission)
- 30 percent margin should be *added* for power estimates (except for EP as noted above)

2.5 Baseline System Design

Due to the large gravity well of Saturn (almost 60 times that of Earth), spiraling down from a high orbit near the sphere of influence would require long periods of time (>5 yr) before approaching Titan. Alternatively, studies have shown that an aerocapture system can place a S/C directly into Titan orbit (Refs. 4, 5, and 6) with no trip time penalties using Titan's atmosphere. Thus the baseline approach was to use an aerocapture system to place an REP S/C in low Titan orbit. The REP S/C is stored inside the aeroshell and goes into operation after aerocapture and the separation of the aeroshell. Options do exist where the REP can be used to reduce the Earth to Saturn transit time.

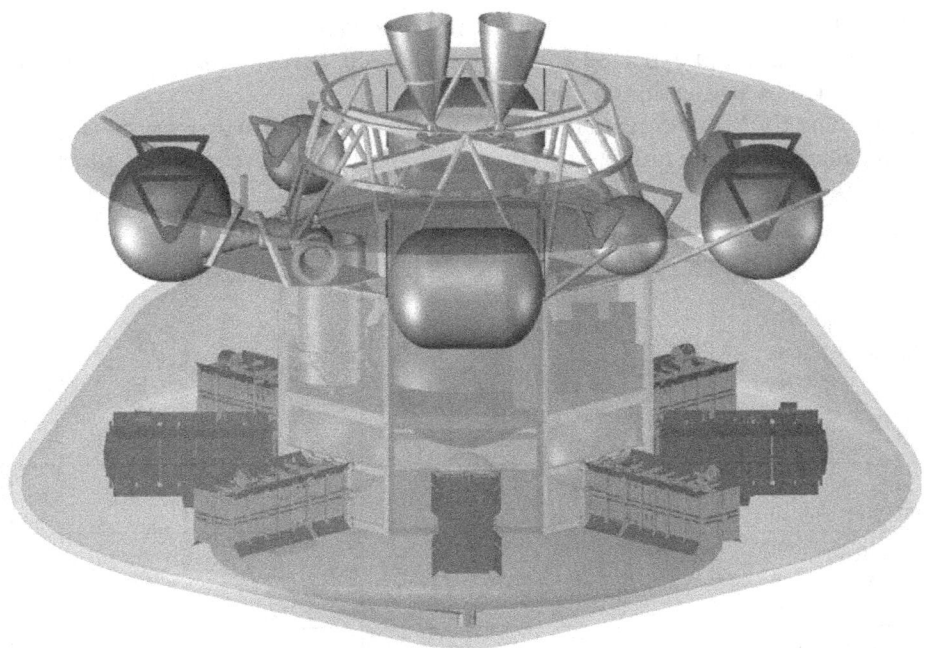

Figure 2.1.—Baseline S/C design interior of aeroshell and cruise deck highlighted.

Figure 2.1 shows the baseline design of the entire S/C. The subsystems on the cruise deck (stage) and the REP S/C are shown opaque. The aeroshell and the structure/radiator of the cruise deck are transparent. The S/C is divided into three stages. The stack consists of a chemical cruise deck, an aeroshell and the REP stage for use in the Saturn Moon system. The chemical cruise deck contains the propulsion to provide a deep space chemical maneuver as well as necessary thermal, navigation, and communications systems for the REP S/C in the aeroshell during the Earth to Saturn transit.

2.6   Mission Description

The REP Flagship mission involved the use of three different propulsion systems: chemical, aerocapture, and electric propulsion, to deliver the REP S/C to the Saturn Moon system. The mission is broken up along the following mission stages

- The REP S/C is launched on a Delta IV H to a C_3 of 48 km^2/s^2.
- The cruise deck chemical stage provides a deep space maneuver with its bi-propellant system.
- The chemical trajectory uses one Earth gravity assist (EGA) on the way to Saturn.
- Upon arrival at Titan, an Aerocapture system enables the S/C to capture into a low Titan orbit using the local Titan atmosphere.
- The REP powered Long Life Hall thruster EP Stage performs drag makeup at Titan, and then spirals the vehicle out of Titan orbit and then down to an orbit about the moon Enceladus (~5 yr trip time).

2.6.1   Interplanetary Chemical Cruise Stage Trajectory Baseline

After performing an initial trade between the uses of the main REP system versus a chemical cruise stage to perform the burn from the Earth to Saturn (Figure 2.2), a chemical cruise stage was chosen as the first propulsion system due to concerns operating the electric propulsion system while inside the aeroshell.

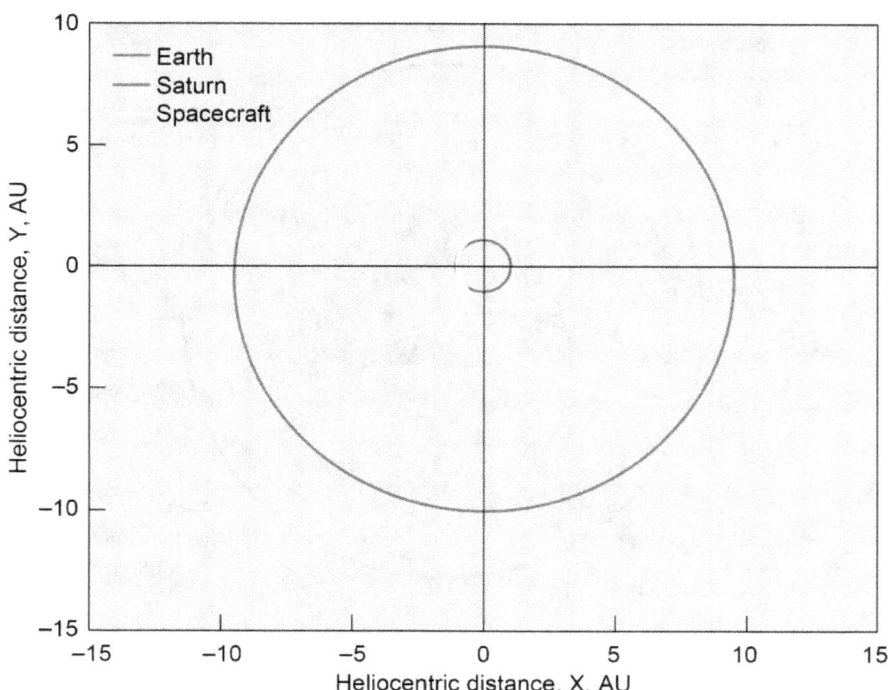

Figure 2.2.—In-space trajectory from Earth to Saturn showing EGA.

Based on previous aerocapture studies for Titan (Refs. 4, 5, and 6), the incoming velocity of the arrival trajectory at Saturn can be as high as 6.5 km/s for the current aeroshell design. Therefore, an Earth-Saturn trajectory was chosen such that the V_{mp} (hyperbolic excess velocity), reported below, was less than that target 6.5 km/s limit.

- Launch vehicle   Delta IV Heavy
- Launch date   March 25, 2015
- Maneuver   October 9, 2016
- Earth fly-by   January 30, 2018
- Arrival at Saturn date   March 23, 2023
- Transfer time   8 yr
- Launch mass   3087 kg (Delta IV-H)
- C_3 48.36 km^2/s^2
- Post launch ΔV 0.699 km/s
- V_{mp} Titan   5.99 km/s (less than the ~ 6.5 entry limit)

2.6.1.1   Aerocapture Maneuver at Titan

The aeroshell will be used to aerocapture into Titan's orbit using Titan's upper atmosphere, recently probed by Huygens. Aerocapturing will eliminate the need to start the REP system at the edge of Saturn's gravity well, thereby significantly reducing mission time. The Aerocapture system is sized based on the 2003 NASA Aerocapture studies (Refs. 4 and 5). The aerocapture maneuver and a subsequent periapsis burn by a dedicated bipropellant thruster system place the S/C in a 1200 km altitude circular orbit. The Aerocapture concept is illustrated in Figure 2.3.

Titan Arrival Orbit Characteristics

- Arrival V_{hp}:   5.99 km/s

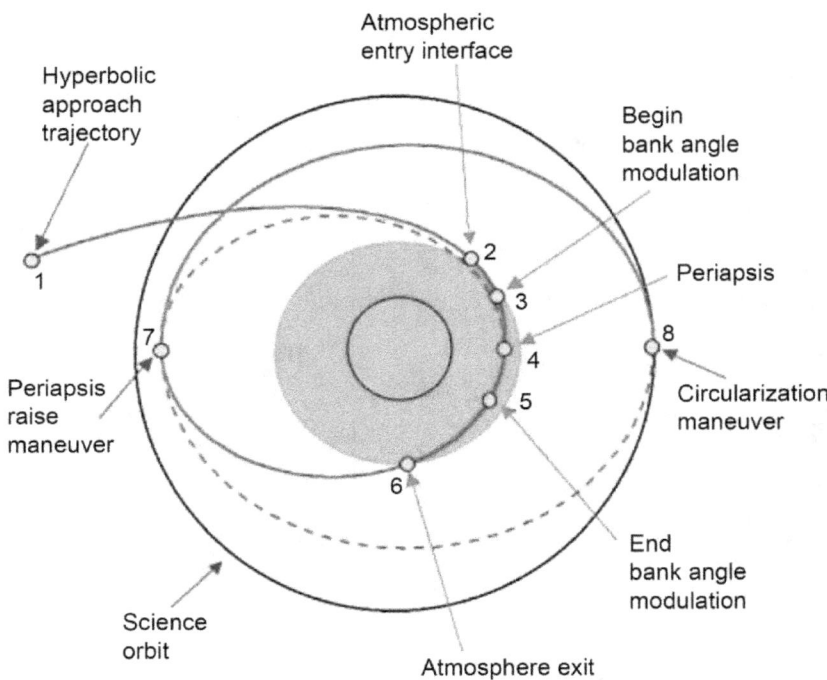

Figure 2.3.—Concept of aerocapture maneuver at Titan.

Titan bipropellant thruster system maneuver budget

- Aerocapture maneuver: 30 m/s
- Periapse raise: 200 m/s
- Attitude control for rest of mission: 50 m/s

Table 2.2 contains the parameters for modeling the aerocapture shell with regard to the Titan orbiter with resulting aerocapture data from the Titan Aerocapture mission (Ref. 7). Table 2.2 also contains the data from the Titan Aerocapture mission as well as similar Mars aerocapture data for comparison.

TABLE 2.2.—DESTINATION PARAMETERS FOR MARS AND TITAN ORBITS (REF. 7)

Destination/reference parameter	Titan	Mars
Entry velocity (km/s)	[a]6.5	5.7
Nominal entry flight path angle (deg)	−36	−14.2
Apoapsis/science orbit (km)	1700	1400
Atmosphere composition	[b]95% N_2, 5% CH_4	95.3% CO_2, 2.7% N_2
Atmosphere scale height at aerocapture altitude (km)	~40	10.5
Atmospheric interface altitude (km)	1000	250
Aerocapture altitude (km)	200 to 400	40
L/D	0.25	0.25
M/CDA (kg/m^2)	[c]90	148
Theoretical corridor (deg)	3.5	~1.4
Time from atmosphere entry to atmosphere exit (min)	42	10
Convective stagnation point heat rate (W/cm^2)	46 (0.91 m nose rad)	30 (1.9 m nose rad)
Radiative aeroheating rate (W/cm^2)	93 to 280	Negligible
Max g's during aerocapture (Earth g's)	3.5	2.5 to 3

Representative aerocaptures at Titan compared to Mars.
[a]Titan aerocapture entry velocity 6.5 to 10 km/sec, comparison given for 6.5 km/sec.
[b]Max CH_4 atmosphere.
[c]For design trajectory and comparison; range: 56 to 84 kg/m^2 dependent on aeroheating, TPS, vehicle diameter

2.6.1.2   Spiral Trajectory Operations at Saturn

The Radioisotope powered electric propulsion system is used to perform drag makeup for the orbit about Titan and to perform the spiral between the Saturn moons Titan and Enceladus. The REP S/C spends 1 yr mapping Titan at a 1200 km apoapse orbit (note the data from the table in Figure 2.2 quotes a higher apoapse than the one applied in this mission, but the data is sufficient for the purpose of this analysis). The Altitude Control System (ACS) thrusters on the REP stage perform orbit maintenance (ΔV of 100 m/s). After the year of science mapping, the REP stage spirals out of Titan orbit and spirals down to Enceladus. The REP stage then spends 1 yr at Enceladus for science mapping. The total ΔV performed by the REP system is approximately 7 km/s at Enceladus. Stability of the REP spiral trajectory due to multibody affect needs to be assessed.

The aerocapture maneuver places the S/C in a 1200 km altitude circular orbit. The S/C will not spiral to Titan, though it may need the REP engine for some orbit maintenance at the low altitude. The spiral from Titan will take an estimated 1836 days. The transfer orbit was calculated using the Edelbaum equation (Refs. 8 and 9) and EP propellant flow. Using the Edelbaum methodology makes the delivered mass only dependent on I_{sp}, and transfer time are dependent on power. For an acceptable trip, the specific mission parameters chosen were

- Transfer from Titan to Enceladus
  - Titan to Enceladus spiral time: 1836 days
  - Specific impulse: 1500 s
  - Thruster input power: 1000 W
  - ΔV: 7.054 km/s
  - Used propellant: 586 kg
  - Propellant with margin: 637 kg

Figure 2.4 shows the delivered mass as a function of trip time and I_{sp} (bottom axis) for the vehicle operating in the Saturn moon system. For the Titan to Enceladus spiral transfer, both trip time and mass fraction (propellant mass/delivered mass) are strongly dependant on specific impulse.

Spiral trades are only for an Edelbaum approximation (assumes 7.054 km/s ΔV regardless of specific impulse). There may be significant gravity losses not taken into account in this analysis. Propellant reserve for trajectory margin is 5 percent. Propellant reserve for residuals is 3.6 percent. Table 2.3 lists the assumptions associated with the low thrust mission design.

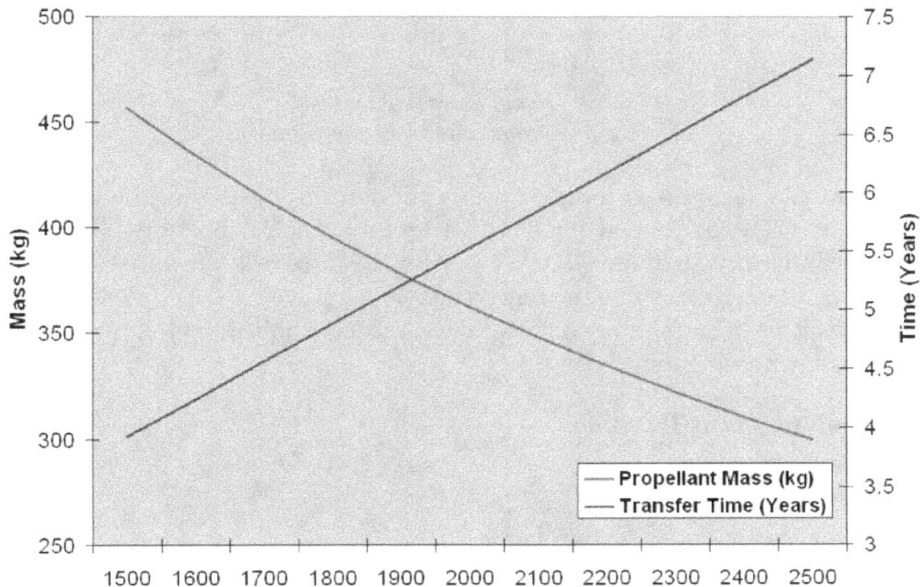

Figure 2.4.—Propellant mass and transfer time as a function of thruster operating I_{sp} (s).

TABLE 2.3.—LOW THRUST MISSION DESIGN ASSUMPTIONS

Mass, Xe total	636.57 kg
Mass, Xe useable	586.1566414 kg
Mass, Xe navigation and trajectory margin	5%
Mass, Xe residuals	3.6%
Specific impulse	1500 s
Power, into the thruster	1000 W
Time, transfer time	2920 days
Time, spiral time	1836 days
ΔV, deep space maneuver	0.7 km/s
ΔV, chemical margin	0.035 km/s
ΔV, spiral	7.05 km/s
Date, launch	March 25, 2015
Date, arrival	March 23, 2023
Mass, launch mass	3087.00 kg
Energy, C_3	48.36 km^2/s^2
Target, flyby	Titan
Target, Vhp	5.99 km/s

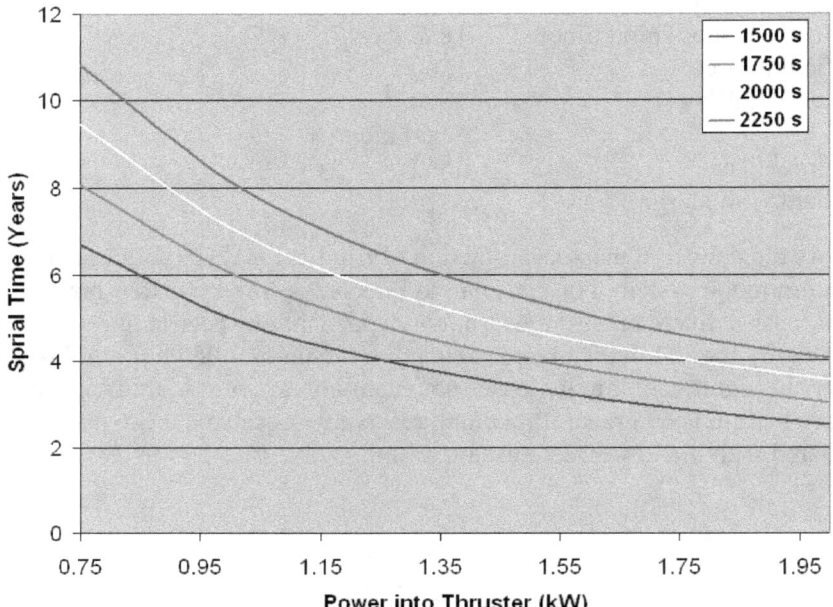

Figure 2.5.—Power sensitivity (spiral phase).

The low thrust mission analysis performed for this mission was done using the low thrust optimization code VARITOP, developed at the NASA Jet Propulsion Laboratory (JPL). In order to simulate the 90 percent duty cycle of the electric thrusters, the thruster level was set to 90 percent. This interjected 10 percent coasting times into the trajectory.

The sensitivity analysis of spiral time to power into thrusters applied the following thruster performance curves shown in Figure 2.5.

2.6.2 Mission Analysis Event Timeline

Main events during transfer
- Launch date: March 25, 2015
- Deep space maneuver: October 9, 2016
- Earth fly-by (EGA):   January 30, 2018
- Arrival date: March 23, 2023

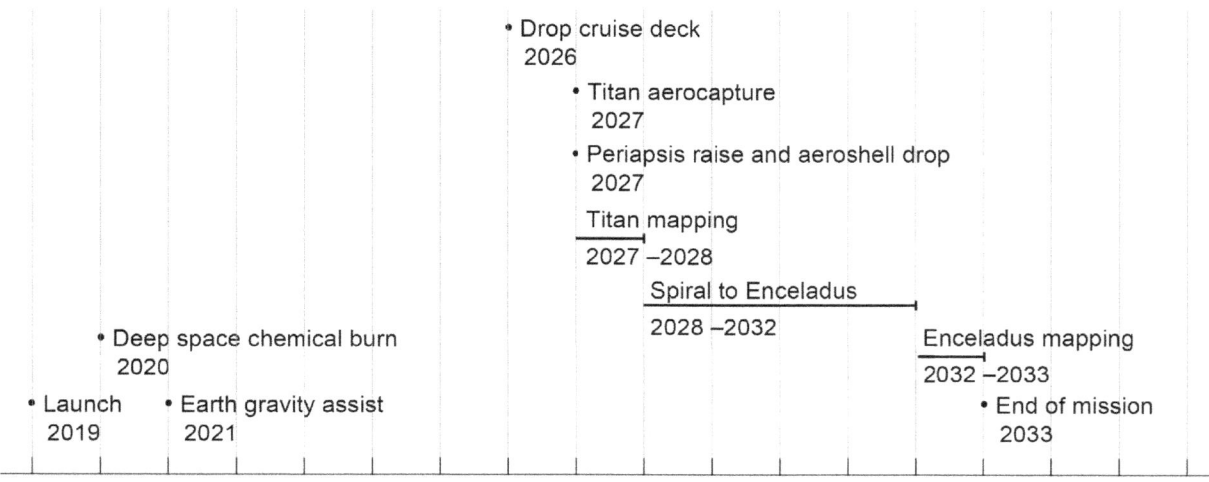

Figure 2.6.—Mission event timeline.

Figure 2.6 is the straw man event timeline for the mission. All dates are relative to the launch date chosen in this mission analysis, and subject to change for different launch date configurations.

Table 2.4 captures the mission event timeline in terms of mass dropping as burns are completed throughout the mission, and as major portions of the S/C are jettisoned. The masses are then used as the initial masses for the next maneuver as the propellant budget is sized via the rocket equation.

TABLE 2.4.—MASS JETTISON HISTORY

Mass jettisoning	Mass (kg)
Launch wet mass	3038.0
Cruise propellant (deep space maneuver)	623.94
S/C at Saturn orbit entry	2414.0
Cruise Deck dry Mass (bottoms-up growth)	311.2
Cruise Deck dry Mass (with system growth)	340.0
Residual prop	18.7
S/C at Saturn post drop Cruise Deck	2055.4
Aeroshell-heat shield (dropped) total	228.2
Aeroshell-backshell (dropped) total	86.4
Aeroshell total mass (bottoms-up growth)	314.6
Aeroshell total mass (with system growth)	342.5
Propellant with aeroshell	27.30
S/C at Saturn post drop aeroshell	1685.6
S/C at Saturn post periapse raise	1541.9

2.7   Launch Vehicle Details

The baseline Launch Vehicle is the Delta IV H. An Atlas 551 with a Star 48 solid propellant stage was considered but lacked the performance required once the bottoms-up REP/Aerocapture/Cruise Deck S/C was designed. The Launch Vehicle performance for launch C_3 is shown in Figure 2.7. This data was gathered through the NASA Kennedy Space Center (KSC) Launch Services website.

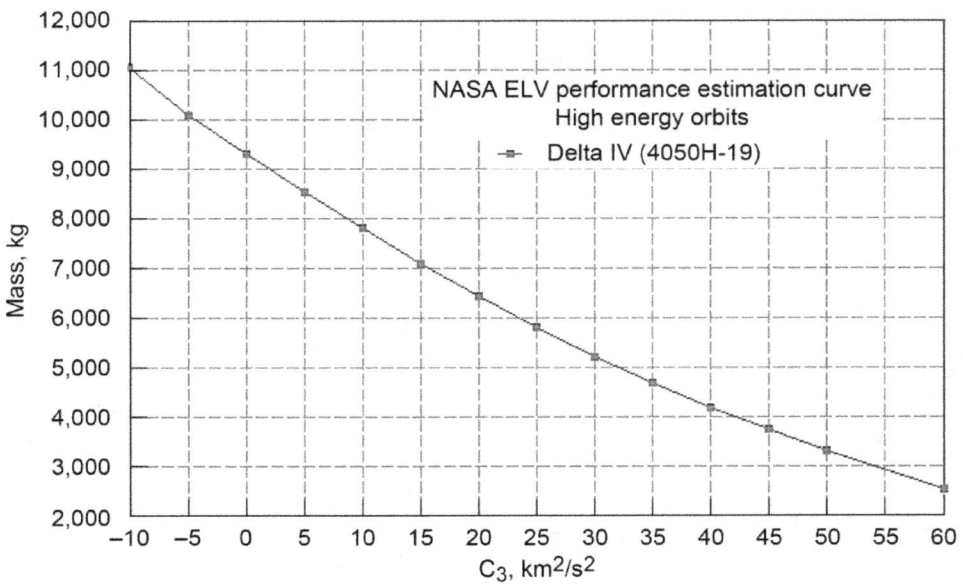

Figure 2.7.—Delta IV H launch mass versus C_3.

Figure 2.8.—REP Flagship
S/C in Delta V payload
fairing.

Figure 2.8 shows the packaging of the REP S/C in the Delta IV H payload fairing. Note that the Delta IV payload shroud volume more than accommodates the 4 m diameter aeroshell thus allowing for options of a 5 m aeroshell to allow more room for the REP S/C and science payload. Options also may exist for adding Titan and Enceladus probes.

2.8   Science Instruments Overview

Typically, the science payload mass delivered by the vehicle in a COMPASS design session is the figure of merit (FOM) of the analysis. In this design, the science payload capability was used as a FOM and was traded off as mass in order to fit the REP vehicle into the launch performance of the Delta IV H. As a starting assumption, a 70 kg science package payload, with 30 percent growth on it was assumed. Additional margin available in launch margin could be traded off for science payload, but would require additional propellant. This is left for future work.

2.9   System Design Configuration Details

The REP S/C will be launched on a Delta IV H. The payload will be located on the inside of the aeroshell and not used until after the aeroshell separates at Titan. To first order, the S/C configuration is built around the following major components:

2.9.1   Cruise Deck Configuration

The Cruise Deck is the name for the stage in this configuration that provides thermal protection and communications, and performs the deep space burn for the cruise to Saturn. It is separated just before aerocapture at Titan. The Cruise Deck includes: radiators to cool the REP S/C inside the aeroshell (including the ASRGs), a propulsion system to provide a single deep-space burn (ΔV of ~700 m/s), a medium gain antenna (MGA), all necessary structure and separation mechanisms, and miscellaneous avionics, cabling, thermal system.

2.9.2   Aeroshell Configuration

The Aeroshell consists of a Heat Shield and a Back Shell. The Aeroshell is dropped after aerocapture into Titan's atmosphere. Aerocapture system is sized on the 2003 NASA Aerocapture studies (Refs. 4, 5, and 6).

2.9.3   REP S/C Configuration

The REP S/C is based on an octagonal frame with a propulsion/power deck and an upper avionics/ communications/science deck. The REP S/C is installed inverted in the aeroshell. The ~600 kg of Xe is carried in a single spherical Composite Overwrapped Pressure Vessel (COPV) tank with two hydrazine monoprop tanks on either side carrying the ACS/periapsis raise propellant. Four advanced Hall thrusters are mounted on the avionics deck since the shape of the aeroshell necessitated placing the 2.1 m fixed antenna below the propulsion deck. The eight ASRGs are spaced radially out from the propulsion deck to allow sufficient clearance with the aeroshell and maximize radiative cooling. The science instruments share the top of the avionics deck with the thrusters. Future assessments will need to consider the proximity of the instruments with the thrusters. The science instruments shown in the REP S/C are from a previous study (Ref. 3) and only representative of the final science payload. Further work is needed to integrate the actual science instrument set.

2.10   Internal COMPASS Details

COMPASS is a multidisciplinary collaborative engineering team whose primary purpose is to perform integrated vehicle systems analysis and provide trades and designs for both Exploration and Space Science Missions.

2.10.1 GLIDE Study Share

GLIDE (GLobal Integrated Design Environment) is a data collaboration tool that enables secure transfer of data between a virtually unlimited number of sites from anywhere in the world. GLIDE is the primary tool used by the COMPASS design team to pass data real-time between subsystem leads.

The study share for this study is: https://glide.grc.nasa.gov/REP_Sept2007

2.10.2 GLIDE Study Container (Architecture)

The convention is to keep the name of the study container and the study share the same. For the COMPASS team, the Microsoft Excel add-in GLIDE Options tool bar pick for the Study Container (Architecture) is:

Study Container (Architecture):   REP_Sept2007

2.10.3 GLIDE Study Container(s)

Reduced Science Case: REP_Flagship_Reducedscience

Study Description

Flagship class mission to the Saturn moon system. Baseline Design (six ASRG). A reduction in the science package for this iteration.

2.11 Top Level Design (MEL and PEL)

The Flagship mission to the Saturn moon system was divided into two distinct stages. To simplify the systems analysis and the dropping of mass during the mission phases.

2.11.1 Master Equipment List (MEL)

2.11.1.1 Chemical Cruise Stage MEL

Table 2.5 lists the top level subsystems of the MEL of the Chemical Cruise Stage.

TABLE 2.5.—CRUISE DECK (STAGE) MEL

WBS no.	Description Generic Chemical Stage	Qty	Unit mass (kg)	CBE mass (kg)	Growth (%)	Growth (kg)	Total mass (kg)
03	Cruise Deck/Chemical Stage	-	-	896.63	6.38	57.23	953.86
03.1	Attitude Determination and Control	-	-	6.46	20.00	1.29	7.75
03.2	Command and Data Handling	-	-	3.00	30.00	0.90	3.90
03.3	Communications and Tracking	-	-	2.40	30.00	0.72	3.12
03.4	Electrical Power Subsystem	-	-	3.00	50.00	1.50	4.50
03.5	Thermal Control (Non-Propellant)	-	-	48.93	15.00	7.34	56.27
03.6	Propulsion	-	-	74.38	30.00	22.31	96.69
03.7	Propellant (Chemical)	-	-	642.63	0.00	0.00	642.63
03.8	Structures and Mechanisms	-	-	115.83	20.00	23.17	139.00

2.11.1.2 REP and Aeroshell MEL

Table 2.6 lists the top level subsystems of the MEL of the REP Stage and the Aeroshell.

TABLE 2.6.—REP AND AEROSHELL MEL

WBS no.	Description REP Flagship Mission Aeroshell (November 7, 2007)	Qty	Unit mass (kg)	CBE mass (kg)	Growth (%)	Growth (kg)	Total mass (kg)
01	REP S/C (Payload and Stage)	-	-	1784.15	10.90	194.52	1978.67
01.1	Science Payload	-	-	70.00	30.00	21.00	91.00
01.2	REP Bus	-	-	1714.15	10.12	173.52	1887.67
01.2.1	Attitude Determination and Control	-	-	18.40	20.00	3.68	22.08
01.2.2	Command and Data Handling	-	-	33.30	34.26	11.41	44.71
01.2.3	Communications and Tracking	-	-	39.00	34.10	13.30	52.30
01.2.4	Electrical Power Subsystem	-	-	223.24	11.51	25.70	248.94
01.2.5	Thermal Control (Non-Propellant)	-	-	61.57	15.00	9.23	70.80
01.2.6	Propulsion	-	-	145.64	28.22	41.10	186.74
01.2.7	Propellant	-	-	847.51	0.00	0.00	847.51
01.2.8	Structures and Mechanisms	-	-	345.49	20.00	69.10	414.59

2.11.2 Power Equipment List (PEL)

The power listing for nominal loads was collected and presented in Table 2.7.

TABLE 2.7.—PEL

	Propulsion (W)	Avionics (W)	Comm. (W)	Thermal (W)	GN&C (W)	Power (W)	Science (W)	CBE total (W)	30 % margin	Total (W)
Launch	0	22	0	33	27	63	0	145	48.3	193
Star 48 operation	0	22	420	33	27	63	0	565	174.3	739
S/C separation	16	22	420	33	27	63	0	581	174.3	755
S/C checkout	16	22	420	33	36	63	265	855	256.59	1112
REP thrusting	1016	22	0	33	29	63	0	1163	48.9	1212
REP coast	16	11	0	33	29	63	2	154	46.23	200
Communications	16	22	420	33	29	63	2	585	175.53	761
Aerocapture	27	22	420	33	29	63	0	594	174.9	769
Titan science	16	22	0	33	29	63	265	428	128.4	556
Spiral to Enceladus	1016	22	0	33	29	63	0	1163	48.9	1212
Enceladus science	16	22	420	33	29	63	265	848	254.4	1102

Thruster input power	950
Thermal waste heat (W)	906
ASRG waste heat (W)	2800
Avionics/comm. or REP (W)	250

2.12 System Level Summary

The S/C mass summary for the cruise deck and REP S/C is shown in Table 2.8. It shows the current best estimate (CBE) and growth masses for each subsystem as well as propellant masses for the propulsion system. The additional system level growth mass to put the total growth at 30 percent is also shown.

TABLE 2.8.—SYSTEM LEVEL SUMMARY WITH GROWTH CALCULATIONS

COMPASS study: Radioisotope Electric Propulsion (REP)				Study date: November 29, 2007	
GLIDE container: *REP_Sept2007: REP_Flagship_reducescience*					
REP S/C MEL rack-up (mass)				COMPASS REP design	
WBS no.	Main subsystems	CBE mass (lkg)	Growth (kg)	Total mass (kg)	Aggregate growth (%)
01	REP S/C (payload and bus)	1784.2	194.5	1978.7	-----
01.1	Science Payload	70.0	21.0	91.0	30.0
01.2	*REP S/C*	*1714.2*	*173.5*	*1887.7*	-----
01.2.1	AD&C	18.4	7.4	22.1	40.0
01.2.2	C&DH	33.3	11.4	44.7	34.3
01.2.3	Communications and tracking	39.0	13.3	52.3	34.1
01.2.4	Electric power	223.2	25.7	248.9	11.5
01.2.5	Thermal control	61.6	9.2	70.8	15.0
01.2.6	Propulsion	145.6	41.1	186.7	28.2
01.2.7	Propellant (Xe and chemical)	847.5	-----	--------	-----
01.2.8	Structures and mechanisms	345.5	69.1	414.6	20.0
	Estimated REP S/C dry mass	937	195	1131.2	20.8
	Estimated REP S/C wet mass	1784	195	1978.7	-----
03	Cruise Deck chemical stage	896.6	57.2	953.9	-----
03.1	AD&C	6.5	1.3	7.8	20.0
03.2	C&DH	3.0	0.9	3.9	30.0
03.3	Communications and tracking	2.4	0.7	3.1	30.0
03.4	Electrical Power subsystem	3.0	1.5	4.5	50.0
03.5	Thermal control (non-propellant)	48.9	7.3	56.3	15.0
03.6	Propulsion	74.4	22.3	96.7	30.0
03.7	Propellant (chemical)	642.6	0.0	642.6	0.0
03.8	Structures and mechanisms	115.8	23.2	139.0	20.0
	Estimated cruise deck dry mass	254	57	311.2	22.5
	Estimated cruise deck wet mass	897	57	953.9	-----
	Total estimated dry mass and wet mass				Total growth
	Estimated Flagship S/C total dry mass	1191	252	1442.4	21.1
	Estimated Flagship S/C total wet mass	2681	252	2932.5	----
	System level growth calculations				Total growth
	Desired system level growth	1191	357	1547.8	30.0
	Additional growth (carried at system level)	-------	105	--------	8.9
	Total wet mass with growth	2681	357	3038.0	
	Available launch performance to C_3 (kg)			3087.0	
	Launch margin available (kg)			49.0	

2.13   Design Concept Drawing and Description

Figure 2.9 shows a side view of the REP Titan/Enceladus Orbiter S/C. All dimensions are in metric units.

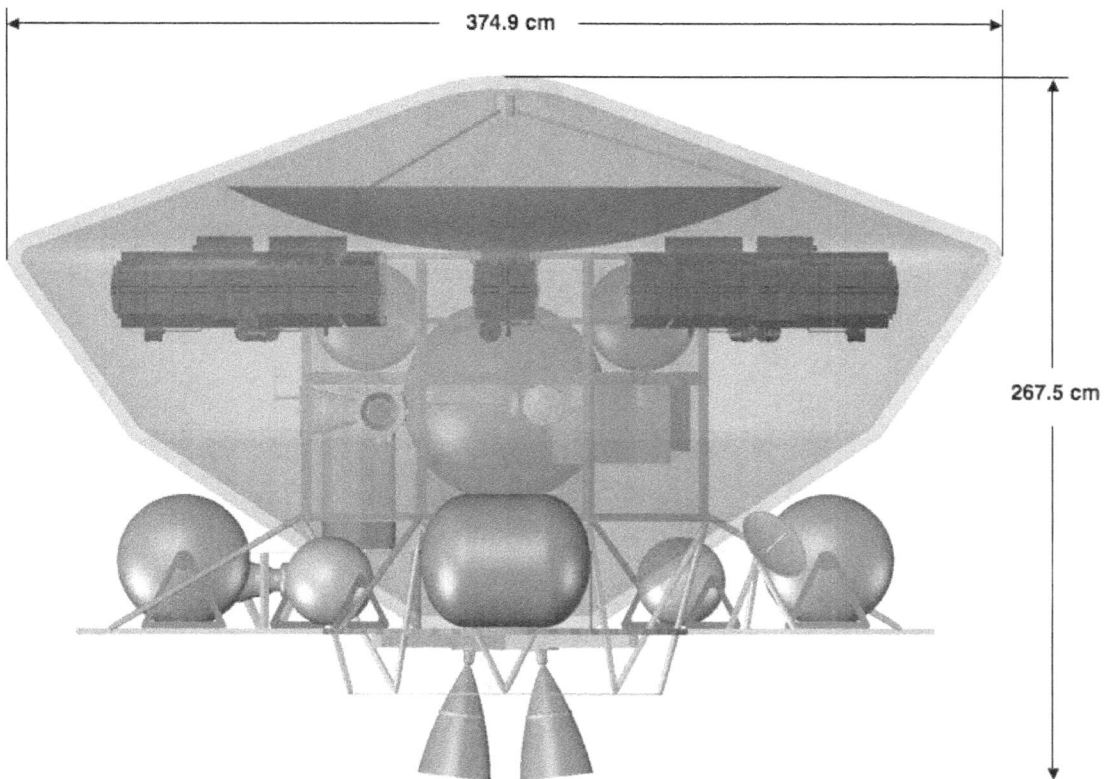

Figure 2.9.—REP stage inside aeroshell with cruise deck attached, with dimensions.

3.0 Subsystem Breakdown

3.1 Attitude Control System (ACS)

The starting design for the ACS on the REP vehicle is borrowed from New Horizons (NH)

- ACS hydrazine
  - Off-the-shelf (OTS) blow-down similar to NH
  - Single tank with ~20 kg hydrazine

3.1.1 ACS Requirements

Autonomous operations, EP navigation approach. Provides full 6-degrees of freedom (DOF) control of the vehicle from separation from the Delta IV H through end of mission.

3.1.2 ACS Assumptions

Much of the design is based on current hardware.

- IMU has Deep Impact, MESSENGER, Cassini, and NEAR heritage
- Sun sensors and Star Trackers taken from IIE Team X design

3.1.3 ACS Design and MEL

Figure 3.1 is the conceptual block diagram of the attitude control system. While the propulsion system (ACS) is included in this block diagram, its mass (hardware and propellant) is being tracked in the propulsion system section of the MEL. Likewise, the instrumentation and computational hardware being used to compute the attitude control information is being kept in the C&DH (avionics) section of the MEL.

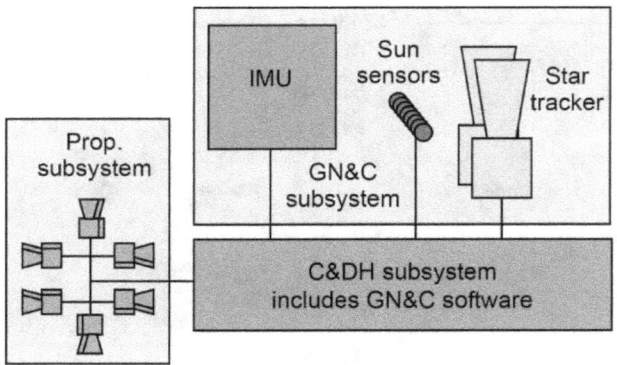

Figure 3.1.—Block diagram of GN&C system.

Figure 3.2.—Northrop Grumman Scalable Inertial Reference Unit (NG SIRU) for space.

Figure 3.3.—Adcole Star Tracker.

3.1.3.1   ACS Model Summary

- One internally redundant IMU (NG SIRU) shown in Figure 3.2.
- Two Star Trackers on the cruise deck and two on the aeroshell/REP S/C (Adcole Corporation) Figure 3.3. These star trackers were the ones used on the NH S/C (Ref. 10)
- Eight Sun Sensors on the cruise deck and eight on the aeroshell/REP S/C (EDO Corp Barnes Engineering Division)
- Four Reaction Wheels (Valley Forge Bearcat 5 Nms reaction wheel, http://www.vfct.com/aerospace/wheels/wheels2.html)
- GN&C software run on main C&DH computers

Table 3.1 lists the items in the ACS MEL for the COMPASS REP S/C design. Table 3.2 lists the items in the ACS MEL for the COMPASS REP/Aeroshell Stage design. All growth allowances follow the AIAA WGA schedule in Section 2.4. Figure 3.4 shows the avionics deck of the REP S/C.

TABLE 3.1.—ACS MEL FOR CHEMICAL CRUISE DECK/STAGE

WBS no.	Description General Chemical Stage	Qty	Unit mass (kg)	CBE mass (kg)	Growth (%)	Growth (kg)	Total mass (kg)
03	Cruise Deck/Chemical Stage	-	----	896.63	6.38	57.23	953.86
03.1	AD&C	-	----	6.46	20.00	1.29	7.75
03.1.1	GN&C	-	----	6.46	20.00	1.29	7.75
03.1.1.a	Sun Sensors	8	0.01	0.08	20.00	0.02	0.10
03.1.1.b	Reaction Wheels	0	0.00	0.00	0.00	0.00	0.00
03.1.1.c	Star Trackers	2	3.19	6.38	20.00	1.28	7.66

TABLE 3.2.—ACS MEL FOR REP/AEROSHELL STAGE

WBS no.	Description REP Flagship Mission Aeroshell (November 7, 2007)	Qty	Unit mass (kg)	CBE mass (kg)	Growth (%)	Growth (kg)	Total mass (kg)
01	REP S/C (Payload and Stage)	-	----	1784.15	10.90	194.52	1978.67
01.2	REP Bus	-	----	1714.15	10.12	173.52	1887.67
01.2.1	AD&C	-	----	18.40	20.00	3.68	22.08
01.2.1.a	GN&C	-	----	18.40	20.00	3.68	22.08
01.2.1.a.b	Sun Sensors	8	0.01	0.04	20.00	0.01	0.05
01.2.1.a.b	Reaction Wheels	4	1.27	5.08	20.00	1.02	6.10
01.2.1.a.c	Star Trackers	2	3.19	6.38	20.00	1.28	7.66
01.2.1.a.d	IMU	1	6.90	6.90	20.00	1.38	8.28

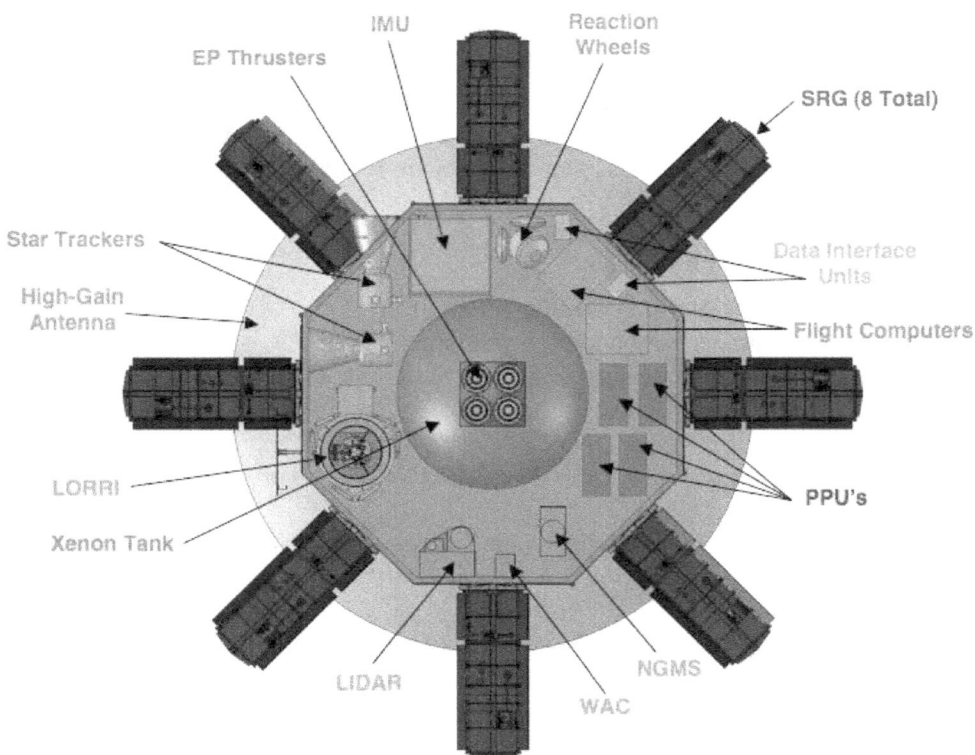

Figure 3.4.—Bottom view of the REP stage highlighting avionics, communications, power, science, and ACS.

3.1.4 ACS Trades

Trades were conducted to use the REP S/C star trackers using openings in the aero backshell. While this is still possible, simplicity drove the addition of two star trackers to the cruise deck.

3.1.5 ACS Analytical Methods

OTS components were used in design.

3.1.6 ACS Risk Inputs

None identified at this time. To be assessed.

3.1.7 ACS Recommendations

Analysis into the amount of ΔV necessary for station keeping and attitude control throughout the mission life needs to be performed to determine whether the 80 m/s assumption is sufficient. Additionally, further research is necessary to determine whether the start trackers and sun sensors are capable of operating at the distances of the Saturnian bodies at the EOL of the trajectory.

3.2 Communications

3.2.1 Communications Requirements

Provide uplink and downlink capability throughout the primary and/or extended mission. The communications system must meet science mission requirements of 8 hr/day of downlink pointed to Earth and a minimum 6.3 kbps downlink at 34-m disk (or about 147 Mbits/day of downlink including a minimum of 10 percent for housekeeping).

3.2.2 Communications Assumptions

Communications requirements are assumed not changed from the REP Centaur orbiter mission in CD–2007–16 (Ref. 3), except more or less input power is available to communications system. Assume the Deep Space Network (DSN) will be capable of supporting Ka-band downlink via 34-m antenna by 2024. The communications system design is based on the NH concept of two onboard-integrated electronics modules (IEM). Overall harness requirements are reduced if the NH IEM design is implemented.

3.2.3 Communications Design and MEL

- REP orbiter communications subsystem consists of
  - A fixed 2.1-m diameter Ka-band high gain antenna (HGA)
  - Two IEMs as in the NH housing many S/C functions, including C&DH, instrument interface circuitry, telemetry interface, solid state recorder, and receiver and exciter sections of the communications subsystem
  - Two 200-W Traveling Wave Tube Amplifier (TWTA) to provide high power RF downlink output
  - RF switch assembly to interconnect antenna with two TWTAs and the rest of communications subsystem
  - Cabling
- Link parameters for Ka-band downlink to 34-m ground stations
  - Ka-band downlink frequency: 32 GHz
  - Range to Earth ground station: 150.0 by 107 km
  - RF power: 200 W = 23.0 dBW
  - Antenna gain: 54.34 dBi (with 50 percent EFF at 32.0 GHz)
  - Pointing loss: 1 dB

- Propagation and polarization loss: 2.0 dB
- Equivalent isotropic radiated power (EIRP): 73.05 dBW
- Cruise Deck
  - Communications system on the cruise stage consists of cabling and a MGA to provide communications while the REP orbiter HGA is inside the aeroshell.

Table 3.3 and Table 3.4 list the items in the communications system MEL for the COMPASS Cruise Deck and the REP/Aerocapture Stage in the REP Flagship S/C design. All growth allowances follow the AIAA WGA schedule in Section 2.4.

TABLE 3.3.—COMMUNICATIONS MEL CHEMICAL CRUISE DECK

WBS no.	Description Generic Chemical Stage	Qty	Unit mass (kg)	CBE mass (kg)	Growth (%)	Growth (kg)	Total mass (kg)
03	Cruise Deck/Chemical Stage	-	----	896.63	6.38	57.23	953.86
03.3	Communications and Tracking	-	----	2.40	30.00	0.72	3.12
03.3.2	MGA	-	----	2.40	30.00	0.72	3.12
03.3.2.a	Transponder	0	0.00	0.00	0.00	0.00	0.00
03.3.2.b	RF Assembly	0	0.00	0.00	0.00	0.00	0.00
03.3.2.c	Processing Module	0	0.00	0.00	0.00	0.00	0.00
03.3.2.d	Antenna	1	2.00	2.00	30.00	0.60	2.60
03.3.2.e	Cabling	2	0.20	0.40	30.00	0.12	0.52

TABLE 3.4.—COMMUNICATIONS MEL REP/AEROCAPTURE STAGE

WBS no.	Description REP Flagship Mission Aeroshell (November 7, 2007)	Qty	Unit mass (kg)	CBE mass (kg)	Growth (%)	Growth (kg)	Total mass (kg)
01	REP S/C (Payload and Stage)	-	----	1784.15	10.90	194.52	1978.67
01.2.3	Communications and Tracking	-	----	39.00	34.10	13.30	52.30
01.2.3.a	X/Ka high gain antenna	-	----	27.00	31.48	8.50	35.50
01.2.3.a.a	Transmitter/receiver	2	4.00	8.00	30.00	2.40	10.40
01.2.3.a.b	Power amp	2	3.00	6.00	30.00	1.80	7.80
01.2.3.a.c	Switch unit	0	0.00	0.00	0.00	0.00	0.00
01.2.3.a.d	Antenna	1	9.00	9.00	30.00	2.70	11.70
01.2.3.a.e	Band pass filter	0	0.00	0.00	0.00	0.00	0.00
01.2.3.a.f	Band reject filter	0	0.00	0.00	0.00	0.00	0.00
01.2.3.a.g	Sensor	0	0.00	0.00	0.00	0.00	0.00
01.2.3.a.h	Cabling	2	2.00	4.00	40.00	1.60	5.60
01.2.3.c	Communications instrumentation	-	----	12.00	40.00	4.80	16.80
01.2.3.c.a	Coaxial Cable	2	6.00	12.00	40.00	4.80	16.80

3.2.4 Communications Trades

With the large power available from the idle EP system, more power than is usually enjoyed by a flagship S/C is available for science and communications. One use of the additional power is to increase the data rate and minimize the DSN contact time (and thus costs).

3.2.5 Communications Analytical Methods

The link budgets provide values of RF transmit powers at 40 and 200 W and antenna gains for Ka-band. Note that X-band is not used in the final design; Ka-band to X-band link budgets are provided for comparison in Table 3.5. Link margins of 3 dB or better exist for all links. The Ka-band with 200 W RF output was baselined to maximize data rate/minimize DSN time. Table 3.5 lists the Communications Link budget analysis performed to size this system.

TABLE 3.5.—LINK BUDGET ANALYSIS

Item	Units	Ka-band downlink		X-band downlink			
Frequency	GHz	32	32	7.75	7.75	7.75	7.75
Transmitter power	W	40	200	40	40	200	200
Transmitter power	dBW	16.0	23.0	16.0	16.0	23.0	23.0
Transmitter line loss	dB	1.0	1.0	1.0	1.0	1.0	1.0
Transmit antenna beamwidth	deg	0.31	0.31	1.29	1.29	1.29	1.29
Peak transmit antenna gain	dBi	54.04	54.04	42.02	42.02	42.02	42.02
Transmit antenna diameter	m	2.1	2.1	2.1	2.1	2.1	2.1
Transmit antenna pointing offset	deg	0.1	0.1	0.37	0.37	0.37	0.37
Transmit antenna pointing loss	dB	1.0	1.0	1.0	1.0	1.0	1.0
Transmit antenna gain (net)	dBi	54.34	54.34	41.72	41.72	41.72	41.72
Equivalent isotropic radiated power	dBW	66.06	73.05	53.74	53.74	60.73	60.73
Propagation path length	km	150×10^7	150×10^7	150×10^7	150×10^7	150×10^7	150×10^7
Space loss	dB	306.04	306.04	293.73	293.73	293.73	293.73
Propagation and polarization loss	dB	2.0	2.0	2.0	2.0	2.0	2.0
Receive antenna diameter	m	34.0	34.0	70.0	34.0	70.0	34.0
Peak receive antenna gain (net)	dBi	78.5	78.5	72.5	66.2	72.49	66.2
Receive antenna beamwidth	deg	0.02	0.02	0.04	0.08	0.04	0.08
Receive antenna pointing error	deg	0.00	0.00	0.00	0.01	0.00	0.01
Receive antenna pointing loss	dB	0.10	0.10	0.10	0.10	0.10	0.10
Receive antenna gain	dBi	78.22	78.22	72.2	65.91	72.2	65.91
System noise temperature	K	15.49	15.49	15.49	15.49	15.49	15.49
Data rate	kbps	7	33	2	0.4	8	2
E_b/N_o	dB	5.10	5.10	5.10	5.10	5.21	5.10
Carrier-to-noise density ratio	dB-Hz	43.29	50.28	37.25	31.0	44.24	37.96
Bit error rate	--------	10^{-5}	10^{-5}	10^{-5}	10^{-5}	10^{-5}	10^{-5}
Required E_b/N_o	dB	2.0	2.0	2.0	2.0	2.0	2.0
Implementation loss	dB	0.10	0.10	0.10	0.10	0.10	0.10
Link margin	dB	3.0	3.0	3.0	3.0	3.11	3.0

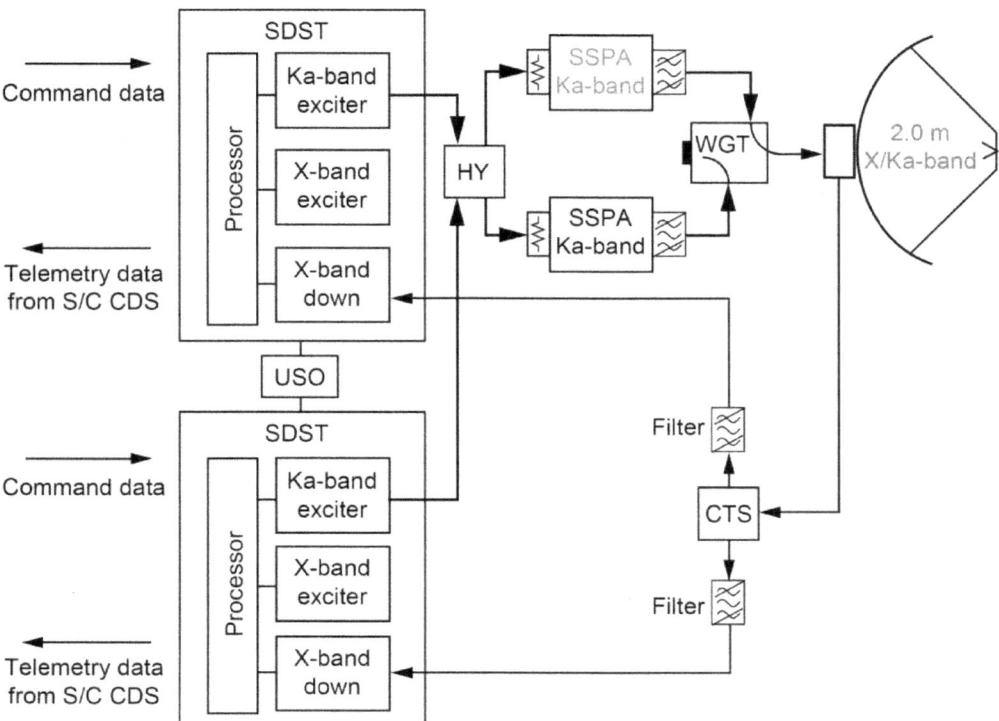

Figure 3.5.—HGA system.

Figure 3.5 is a graphical representation of the communications system of a HGA courtesy of JPL.

3.2.6   Communications Risk Inputs

None assessed at this time.

3.2.7   Communications Recommendation

In the future, further analysis should be done considering the use of the communications system from the NH mission. Figure 3.6 shows a detailed block diagram of that communications system.

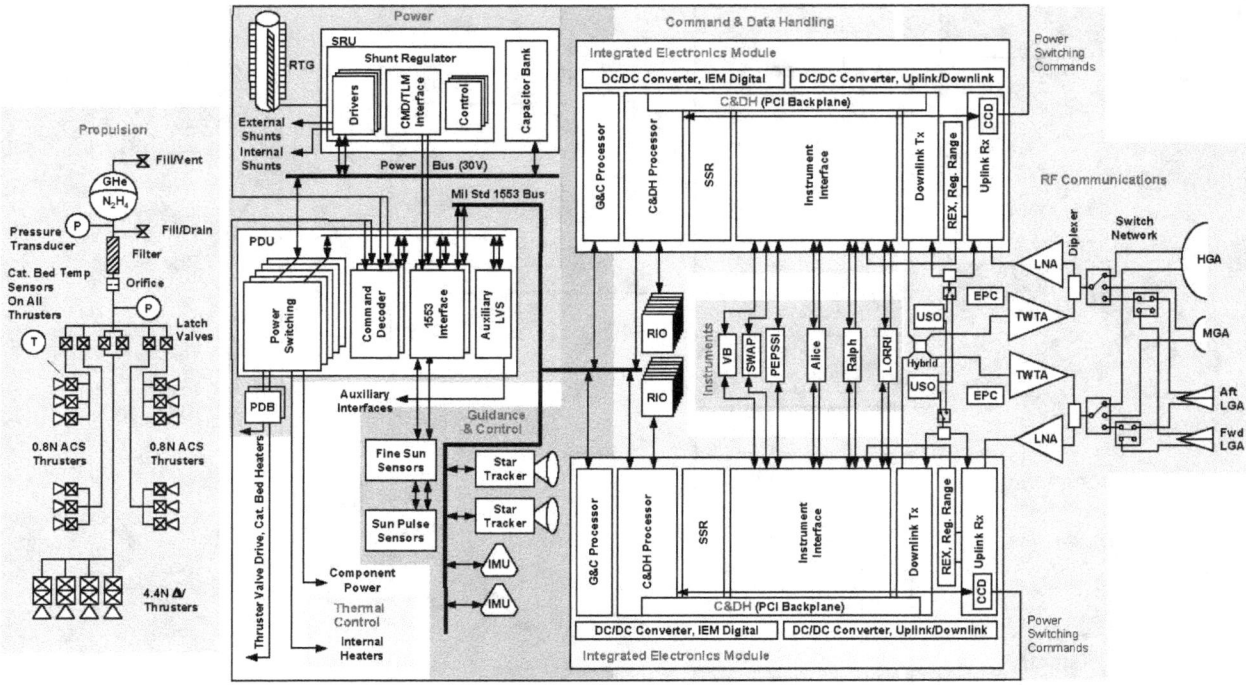

Figure 3.6.—NH avionics layout.

3.3   Command and Data Handling (C&DH)—(Avionics)

3.3.1   Avionics Requirements

The Design requirements, from the science payload and the REP Bus, for the C&DH system were as follows

- Storage for TBD days of data (TBD, est. 8 to 16 GB)
- Avionics for systems command, control, and health management
- Payload control will be done by the C&DH system
- Single fault tolerant avionics

3.3.2   Avionics Assumptions

- All electronics are ≥65 Krad avionics
- Cabling is estimated as 50 percent of the avionics hardware mass
- Avionics spares are cold spares to minimize power consumption
- NH S/C was used as the starting point for the avionics hardware design

3.3.3   Avionics Design and MEL

All avionics components used in the design are based on commercially available components from BAE and SEAKR. There are two independent avionics boxes to provide for single fault tolerance. Each avionics box contains a GN&C/C&DH RAD6000 processor, 256 MB GN&C solid state memory card, SSR card, a Comm. interface card, and a payload interface card. The 1553 processor is used for communications between the GN&C processor and GN&C hardware, i.e., star trackers, IMUs, etc. The GN&C and C&DH computers communicate via the 1553 bus.

Table 3.6 lists the components used in the COMPASS C&DH MEL design on the cruise deck and Table 3.7 lists the C&DH design in the REP/Aeroshell stage. These are the inputs from the subsystem lead. All growth allowances follow the AIAA WGA schedule in Section 2.4.

TABLE 3.6.—AVIONICS (C&DH) MEL CHEMICAL CRUISE DECK

WBS no.	Description Generic Chemical Stage	Qty	Unit mass (kg)	CBE mass (kg)	Growth (%)	Growth (kg)	Total mass (kg)
03	Cruise Deck/Chemical Stage	-	----	896.63	6.38	57.23	953.86
03.2	C&DH	-	----	3.00	30.00	0.90	3.90
03.2.1	C&DH	-	----	3.00	30.00	0.90	3.90
03.2.1.a	Flight computer	0	0.00	0.00	0.00	0.00	0.00
03.2.1.b	Command and telemetry computer	0	0.00	0.00	0.00	0.00	0.00
03.2.1.c	Data interface unit	1	2.00	2.00	30.00	0.60	2.60
03.2.1.d	Data bus operations amplifier	0	0.00	0.00	0.00	0.00	0.00
03.2.1.e	Operations recorder	0	0.00	0.00	0.00	0.00	0.00
03.2.1.f	Command and control harness (data)	1	1.00	1.00	30.00	0.30	1.30

TABLE 3.7.—AVIONICS (C&DH) MEL REP/AEROSHELL STAGE

WBS no.	Description REP Flagship Mission Aeroshell (November 7, 2007)	Qty	Unit mass (kg)	CBE mass (kg)	Growth (%)	Growth (kg)	Total mass (kg)
01	REP S/C (Payload and Stage)	-	------	1784.15	10.90	194.52	1978.67
01.2.2	C&DH	-	------	33.30	34.26	11.41	44.71
01.2.2.a	C&DH	-	------	33.30	34.26	11.41	44.71
01.2.2.a.a	Flight computer	2	8.00	16.00	25.00	4.00	20.00
01.2.2.a.b	Command and telemetry computer	0	0.00	0.00	0.00	0.00	0.00
01.2.2.a.c	Data interface unit	2	2.00	4.00	30.00	1.20	5.20
01.2.2.a.d	Data bus operations amplifier	0	0.00	0.00	0.00	0.00	0.00
01.2.2.a.e	Operations recorder	2	1.10	2.20	30.00	0.66	2.86
01.2.2.a.f	Command and control harness (data)	1	11.10	11.10	50.00	5.55	16.65

3.3.4 Avionics Trades

None performed to date, but comparison to NH hardware to be completed at a later date.

3.3.5 Avionics Analytical Methods

OTS components were used in design.

3.3.6 Avionics Concerns, Comments, Recommendations

- No ultra-stable oscillator (USO)/atomic clock included in avionics hardware. Should it be included in Communication system?
- Processing power of the RAD6000 is assumed to be adequate for GN&C, C&DH, and science payload
- Storage requirements are driven by fly-by storage needs (which are still being estimated)
- Only one SSR would be active at a time and thus susceptible to SEUs
- Total radiation dose is a concern with all deep space missions. This preliminary design has attempted to use only hardware which has already been proven in a deep space mission to assure the life of the electronics over the 12-yr mission.

3.4 Electrical Power System

3.4.1 Power Requirements

Minimize power for non-propulsion during EP operation (minimize plutonium needed).

3.4.2 Power Assumptions

The baseline REP Stage design used eight ASRGs for the generation of power to be used by the science instruments, avionics and electric propulsion thrusters. The specific power assumptions of the ASRGs used are listed under Figure 3.7 in the following section.

3.4.3 Power Design and MEL

Eight ASRGs (12 GPHS) are designed to provide 1120 W to power the REP S/C at beginning of life (BOL). The system is designed to provide 1040 W to the REP S/C at EOL (10-yr). There are negligible thermal interactions between the ASRGs. Figure 3.7 shows a typical ASRG with the main components called out in the graphic. The eight are connected together via a Shunt Regulator/Bus Protection (RBI) assembly. This RBI isolates the ASRGs from S/C bus and each other and follows load demands from S/C bus. There is an approximately 6 percent loss through the RBI and monitoring circuitry (94 percent of power flows through to loads) with 53 W used for fault detection/monitoring. Included in this system is a bus Capacitance of 3000 µf which provides some bus rigidity. Power cabling and harness systems design assumes a 1 percent line loss.

- Specific performance details on the ASRG unit are as follows:
    - Power: 140 W at 28 ± 0.2 V BOL
                         130 W at 28 ± 0.2 V EOL (10 yr)
                         EOM Deep Space (14 yr) 126 We
    - Mass: 20.2 to 21.5 kg with mounting isolator plate
    - Envelope: 30.5 cm W, 46 cm H, 76 cm L (12 in. W, 18 in. H, 30 in. L)
    - Specific power: 6.7 We/kg

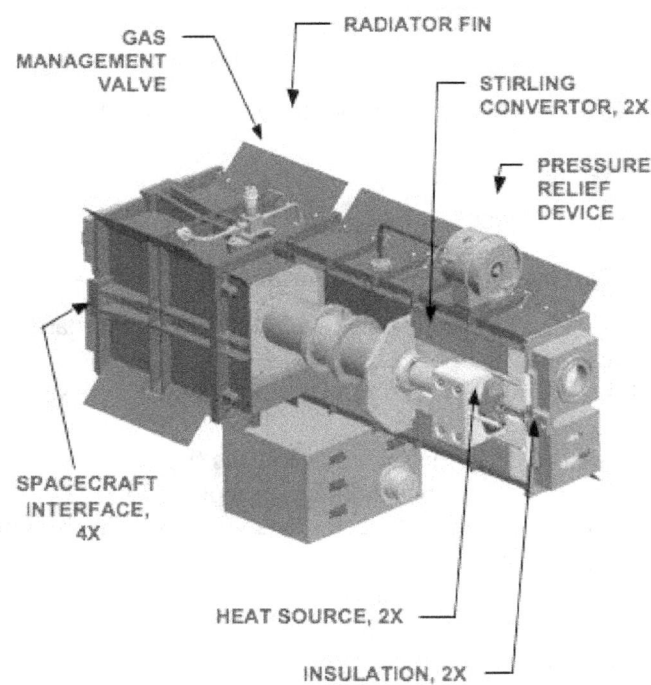

Figure 3.7.—ASRG computer aided design (CAD) model.

ASRG Design Attributes
- Two Stirling converters
  - Co-axially aligned for dynamic balance
  - One GPHS module per converter
- Integrated, single-fault tolerant controller
- Autonomous operation and fault isolation from S/C
- S/C disturbance torque requirement < 35 N-m
  - Based on 1000 kg, 1-m cube S/C with 5-μrad pointing accuracy and a safety factor of 5

Table 3.8 lists the items in the Power system MEL for the Cruise deck design. Table 3.9 lists the items in the Power system MEL for the COMPASS REP/Aeroshell S/C design. All growth allowances follow the AIAA WGA schedule in Section 2.4.

TABLE 3.8.—POWER SYSTEM MEL CHEMICAL CRUISE DECK

WBS no.	Description Generic Chemical Stage	Qty	Unit mass (kg)	CBE mass (kg)	Growth (%)	Growth (kg)	Total mass (kg)
03	Cruise Deck/Chemical Stage	-	----	896.63	6.38	57.23	953.86
03.4	Electrical power subsystem	-	----	3.00	50.00	1.50	4.50
03.4.3	Power cable and harness subsystem	-	----	3.00	50.00	1.50	4.50
03.4.3.a	S/C bus harness	0	0.00	0.00	0.00	0.00	0.00
03.4.3.b	PMAD harness	0	0.00	0.00	0.00	0.00	0.00
03.4.3.c	Electric propulsion harness	0	0.00	0.00	0.00	0.00	0.00
03.4.3.d	RPS to S/C harness	0	0.00	0.00	0.00	0.00	0.00
03.4.3.e	Power cabling	1	3.00	3.00	50.00	1.50	4.50

TABLE 3.9.—POWER SYSTEM MEL REP/AEROSHELL STAGE

WBS no.	Description REP Flagship Mission Aeroshell (November 7, 2007)	Qty	Unit mass (kg)	CBE mass (kg)	Growth (%)	Growth (kg)	Total mass (kg)
01	REP S/C (Payload and Stage)	-	------	1784.15	10.90	194.52	1978.67
01.2.4	Electrical Power Subsystem	-	------	223.24	11.51	25.70	248.94
01.2.4.a	RPS	-	------	155.76	10.00	15.58	171.34
01.2.4.a.a	RPS Main System	8	19.47	155.76	10.00	15.58	171.34
01.2.4.a.b	Miscellaneous no. 2	0	0.00	0.00	0.00	0.00	0.00
01.2.4.b	PMAD	-	------	32.48	15.00	4.87	37.35
01.2.4.b.a	Power management/control electronics	0	0.00	0.00	0.00	0.00	0.00
01.2.4.b.b	Power distribution/monitoring wiring harness	0	0.00	0.00	0.00	0.00	0.00
01.2.4.b.c	DC switchgear/shunt regulator	1	32.48	32.48	15.00	4.87	37.35
01.2.4.b.d	Miscellaneous no. 2	0	0.00	0.00	0.00	0.00	0.00
01.2.4.c	Power cable and harness subsystem	-	------	35.00	15.00	5.25	40.25
01.2.4.c.a	S/C bus harness	1	7.00	7.00	15.00	1.05	8.05
01.2.4.c.b	PMAD harness	1	7.00	7.00	15.00	1.05	8.05
01.2.4.c.c	Electric propulsion harness	1	7.00	7.00	15.00	1.05	8.05
01.2.4.c.d	RPS to S/C harness	1	7.00	7.00	15.00	1.05	8.05
01.2.4.c.e	Power cabling	1	7.00	7.00	15.00	1.05	8.05

3.4.4 Power Trades

A further trade will be made on the number of RTGs necessary to provide sufficient power to the thrusters and instruments to perform the mission.

For the power system, two other options under consideration were as follows

Option 1

- Direct drive the Hall thrusters
- Use of dual alternator (providing 600 V and 28 V, 100 Hz AC)
- 10/1 power ratio on dual alternators
- The 600 V AC converts to 400 V DC
- Power to thrusters EOM 646 W
- Power to payload EOM 76 W

Option 2

- DC/DC conversion to 28 V DC provided from ASRG as designed
- DC/DC conversion to 400 V DC for hall thruster
- The current estimate of single 600 W DC/DC converter at 30 kg

Table 3.10 lists the impact of trade in the number of SRGs and total power available, as well as excess power to be radiated.

- Eight SRGs provide (1120 W BOL, 1040 EOL) 750 W power into the thruster with excess 14 W EOL
- Loss of SRG would limit the power to ~650 W into thrusters

TABLE 3.10.—TRADE ANALYSIS OF VARYING NUMBER OF SRGS

Number SRG	4	5	6	7	8
Power (EOL, 10 yr)	130	130	130	130	130
Total Power EOL (W)	520	650	780	910	1040
Into thruster (W)	250	400	500	650	750
PPU, Line Loss	25	40	50	65	75
Housekeeping (cruise only)	155	155	155	155	155
Housekeep margin (30%)	46.5	46.5	46.5	46.5	46.5
Excess	44	9	29	−7	14

3.4.5 Power Analytical Methods

The modeling of the ASRG units were based on current analysis being done at GRC in the area of ASRG development.

3.4.6 Power Risk Inputs

Will there be power available in current configuration?

- If the angle between ASRGs were changed to 90°, better view
- Vibration/thermal leak to science

3.4.7 Power Recommendation

Further analysis should be done on the alternate power options outlines in Section 3.4.4.

3.5   Structures and Mechanisms

3.5.1   Structures and Mechanisms Requirements

The REP S/C structure must contain necessary hardware for instrumentation, avionics, communications, propulsion and power. It must be able to withstand applied loads from launch vehicle and provide minimum deflections, sufficient stiffness, and vibration damping. This analysis assumed a maximum axial load of 6g. The goal of the design is to minimize weight of the components that make up the structure of the S/C bus, and must fit within the physical confines of the launch vehicle.

3.5.2   Structures and Mechanisms Assumptions

The basic assumptions made in the design process of the S/C bus structure were

- Material: Al alloy 2090-T3
- Space frame with tubular members
- Composite sandwich structure shelf assumed to be all Al using Al 2090-T3 face sheets and an Al honeycomb core with the trade name, Alcore Higrid
- Welded and threaded fastener assembly

3.5.3   Structures and Mechanisms Design and MEL

3.5.3.1   Chemical Cruise Stage

The cruise deck is used to perform the ΔV from Earth escape to Titan capture. A separate chemical cruise stage MEL (Table 3.11) was developed to track the items of the chemical cruise stage. This way, the stage and its mass were more easily dropped during the system modeling and mission modes to accurately calculate propellant loads at each mission phase.

TABLE 3.11.—STRUCTURES AND MECHANISMS MEL CHEMICAL CRUISE STAGE

WBS no.	Description Generic Chemical Stage	Qty	Unit mass (kg)	CBE mass (kg)	Growth (%)	Growth (kg)	Total mass (kg)
03	Cruise Deck/Chemical Stage	-	-------	896.63	6.38	57.23	953.86
03.8	Structures and mechanisms	-	-------	115.83	20.00	23.17	139.00
03.8.1	Structures	-	-------	108.60	20.00	21.72	130.32
03.8.1.a	Primary structures	-	-------	90.42	20.00	18.08	108.50
03.8.1.a.a	Main bus structure	1	10.77	10.77	20.00	2.15	12.92
03.8.1.a.b	Cruise deck structure	1	79.65	79.65	20.00	15.93	95.58
03.8.1.b	Secondary structures	-	-------	18.18	20.00	3.64	21.82
03.8.1.b.a	Balance mass	0	0.00	0.00	0.00	0.00	0.00
03.8.1.b.b	Tank supports and bracketry	1	18.18	18.18	20.00	3.64	21.82
03.8.2	Mechanisms	-	-------	7.23	20.00	1.45	8.68
03.8.2.e	Adaptors and separation	-	-------	7.23	20.00	1.45	8.68
03.8.2.e.b	Separation mechanism from LV	1	2.88	2.88	20.00	0.58	3.46
03.8.2.e.d	Separation mechanism REP probe	1	4.35	4.35	20.00	0.87	5.22

3.5.3.2   Aeroshell

The aeroshell is modeled using the 2003 NASA Titan Aerocapture studies data (Refs. 4, 5, and 6). It can sustain aerocapture maneuvers up to 6.5 km/s. The aeroshell components are modeled in the main REP MEL (see Table 3.12).

Figure 3.8 is the REP stage of the vehicle with dimensions. The antenna is down, the ASRGs point out to left and right. The Aeroshell and Chemical Cruise Deck are not shown.

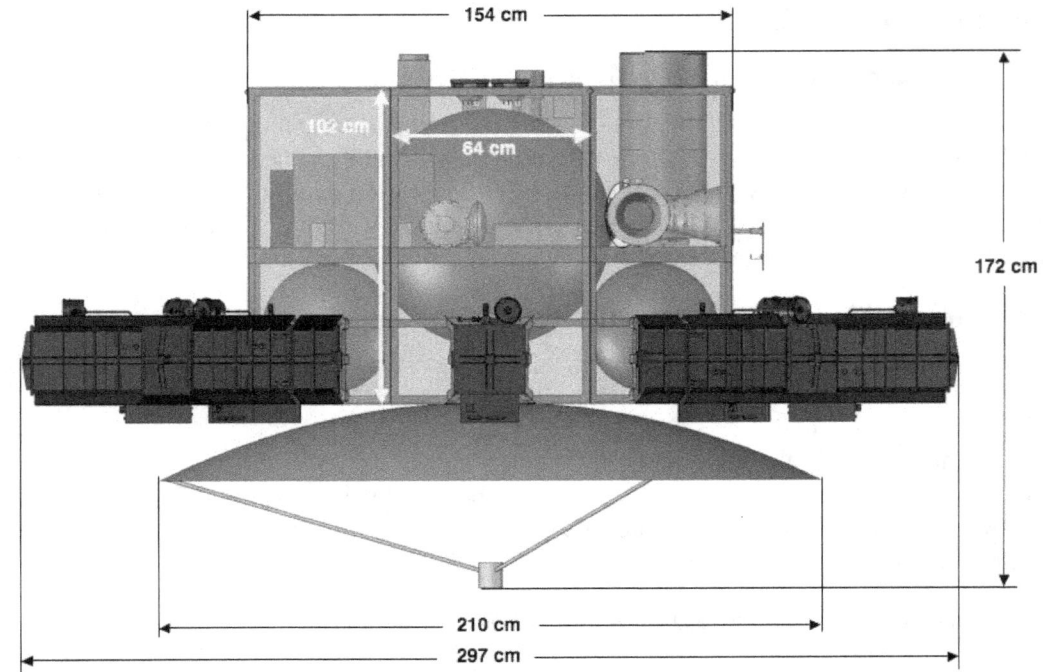

Figure 3.8.—REP S/C with dimensions.

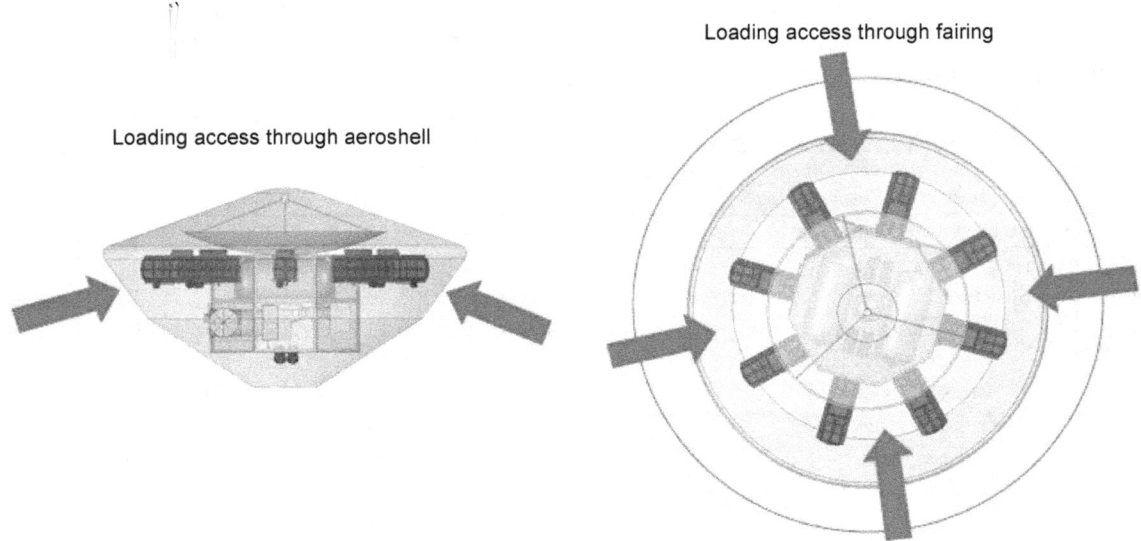

Figure 3.9.—ASRG loading access to REP S/C design.

The structural design of the Aeroshell and the shroud needs to allow for four access panels to the ASRGs while the REP S/C is being loaded on the pad. Figure 3.9 shows the REP S/C within the Aeroshell and within the ELV Fairing, noting where access points would need to be accommodated in the design. The four separate ports can be seen in the top down view on the right.

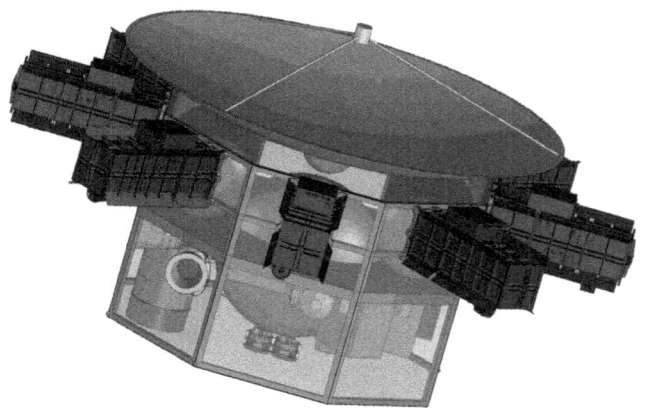

Figure 3.10.—REP S/C (after aeroshell separation).

3.5.3.3 REP S/C

Description of the Design

- Tubular space frame in octagonal configuration
- Shelf of composite sandwich architecture with honeycomb core to mount hardware
  - Composite sandwich structure shelf assumed to be all Al using Al 2090-T3 face sheets and an Al honeycomb core with the trade name, Alcore Higrid.
- Thin sheets for shear panels and to enclose structure
- ASRGs mounted to main bus through vibration isolators

Figure 3.10 gives the conceptual design of the base REP S/C inside the aeroshell. All growth allowances follow the AIAA WGA schedule in Section 2.4.

TABLE 3.12.—STRUCTURES AND MECHANISMS REP STAGE AND AEROSHELL MEL

WBS no.	Description REP Flagship Mission Aeroshell (November 7, 2007)	Qty	Unit mass (kg)	CBE mass (kg)	Growth (%)	Growth (kg)	Total mass (kg)
01	REP S/C (Payload and Stage)	-	--------	1784.15	10.90	194.52	1978.67
01.2.8	Structures and mechanisms	-	--------	345.49	20.00	69.10	414.59
01.2.8.a	Structures	-	--------	330.63	20.00	66.13	396.76
01.2.8.a.a	Primary structures	-	--------	225.83	20.00	45.17	271.00
01.2.8.a.a.a	Main bus structure	1	35.63	35.63	20.00	7.13	42.76
01.2.8.a.a.b	Remove this cruise deck mass	0	0.00	0.00	20.00	0.00	0.00
01.2.8.a.a.c	Aeroshell-heat shield (dropped)	1	190.20	190.20	20.00	38.04	228.24
01.2.8.a.b	Secondary structures	-	--------	104.80	20.00	20.96	125.76
01.2.8.a.b.a	Aeroshell-backshell (dropped)	1	71.96	71.96	20.00	14.39	86.35
01.2.8.a.b.b	Tank supports and bracketry	1	4.55	4.55	20.00	0.91	5.45
01.2.8.a.b.c	SRG support structure	8	2.63	21.03	20.00	4.21	25.23
01.2.8.a.b.d	SRG vibration isolation hardware	8	0.91	7.27	20.00	1.45	8.73
01.2.8.b	Mechanisms	-	--------	14.86	20.00	2.97	17.83
01.2.8.b.f	Installations	-	--------	14.86	20.00	2.97	17.83
01.2.8.b.f.a	Science payload installation	1	1.76	1.76	20.00	0.35	2.12
01.2.8.b.f.b	C&DH installation	1	1.33	1.33	20.00	0.27	1.60
01.2.8.b.f.c	Communications and tracking installation	1	1.56	1.56	20.00	0.31	1.87
01.2.8.b.f.d	GN&C installation	1	0.74	0.74	20.00	0.15	0.88
01.2.8.b.f.e	Electrical Power Installation	0	0.00	0.00	0.00	0.00	0.00
01.2.8.b.f.f	Thermal control installation	1	3.03	3.03	20.00	0.61	3.63
01.2.8.b.f.g	Electric propulsion installation	1	6.44	6.44	20.00	1.29	7.73

3.5.4 Structures and Mechanisms Trades

Next steps in the analysis are to perform a trade on the use of composite for the main bus compartment structure, sizing of space frame to accommodate requirements for antenna, SRGs, and instrumentation while fitting within confines of launch vehicle.

3.5.5 Structures and Mechanisms Analytical Methods

Preliminary structural analysis and modeling was performed using the given launch loads and dimensions of the desired S/C bus. The loads assumptions were 6g axial loading, 3.5g lateral loading (not concurrent with max axial loading), using Al 2090-T3 as the material and a 1.4 safety factor. An additional installation mass was held for each subsystem in the mechanisms section of the structures system. These installations were modeled using 4 percent of the CBE dry mass of each of the subsystems. A 20 percent growth margin was applied to that installation mass.

3.5.6 Structures and Mechanisms Risk Inputs

Risk analysis to be performed.

3.5.7 Structures and Mechanisms Recommendation

Use a more detailed structural analysis for loads and vibrations using a modeling tool, i.e., finite element analysis (FEA). The use of graphite/polymer composites on the Shelf face sheets and outer sheets may provide for further weight savings but at increased cost.

3.6 Propulsion and Propellant Management

3.6.1 Propulsion and Propellant Management Requirements

This Flagship mission combined three different propulsion/capture technologies to perform various stages of the mission in order to deliver the maximum S/C mass or fit inside of the performance cap of the Delta IV H launch vehicle. These three modes were a chemical cruise stage, an aeroshell capture into Titan, and an REP powered EP spiral thrusting system. The two propulsion options: chemical cruise stage and REP moon spiral system will be discussed here.

The Chemical Cruise Stage was required to provide the necessary propulsion burns during the Earth-Saturn Transit stage of the mission. These burns were required to provide necessary ΔV for orbit shaping and necessary ΔV for attitude control. Because these burns are completed during transit to Saturn, S/C mass can be reduced substantially by dropping the no longer needed elements thereby reducing subsequent propulsion requirements later in the mission.

The requirements for the REP system in spiraling from Titan to Enceladus and associated maneuvers are a result of a trade between trip time and payload mass. Trip time is inversely proportional to the I_{sp}, while payload mass will follow the rocket equation. The requirements chosen for the purpose of this study are

- 1000 W
- 1500 s I_{sp}
- 587 kg throughput

3.6.2 Propulsion and Propellant Management Assumptions

3.6.2.1 Chemical Cruise Stage

Both primary and attitude control thrusters are single-fault tolerant as they both have redundant engines in place for their respective burns. However, vehicle operations may be required to enable redundant use of attitude control thrusters. The fuel monomethyl hydrazine (MMH) and oxidizer (nitrogen tetroxide (NT)) are stored in two cylindrical tanks with a He pressurization system. All valving is assumed to be dual-seat to insure required sealing. Propellant lines, tanks, and pressurization are zero-fault tolerant (single string). Propellant tanks and lines are wrapped in tape heaters and insulation to insure that the propellant does not freeze, which could have very detrimental results. Further description of the propulsion system configuration is provided below.

3.6.2.2   REP Stage

The baseline system redundancy assumption was based on single string units. In other words, a propulsion system unit consists of a string of thruster, PPU, gimbal, and propellant management system. Spares or redundant units are assumed to consist of all of the above subsystems. It is important to note that the mission is already modeled with a 90 percent duty cycle. So, 10 percent of the time, the S/C is coasting along its trajectory.

The RCS propulsion subsystem follows the same design assumptions as the chemical propulsion system on the cruise stage

The baseline propulsion system design consists of the following items

- One active 1000 W Long Life Hall engine with two extra for life and one cold spare
- Four PPUs: no cross-strap
- Two-axis range of motion: TBD
- 1 percent Xe unusable
- Pounds per square inch spherical COPV Xe tank
- OTS hydrazine system with NH heritage

3.6.3   Propulsion and Propellant Management Design Trades

3.6.3.1   Chemical Cruise Stage

The only trade study made for the Cruise stage was to compare monoprop versus biprop propulsion option for the Auxiliary Propulsion System.

- For the monoprop option:
  - Auxiliary thruster: Aerojet MR-104 thruster, 441 N, 239 s I_{sp}
  - Reaction thrusters: Aerojet MR-111 thruster, 4.4 N thrust, 229 s I_{sp}
  - Propellant storage configuration: Single fuel (hydrazine) COPV spherical tank with a blowdown pressurization system
- For the biprop option:
  - Auxiliary thruster: Aerojet R-4D "HiPAT" thruster, 445 N thrust, 323 s I_{sp}
  - Reaction thrusters: Rocketdyne R-53 thruster, 8.9 N thrust, 295 s I_{sp}
  - Propellant storage configuration: Dual fuel (MMH) and Oxidizer (NTO) COPV cylindrical tanks with single dedicated He pressurization tank

In this comparison, the Reaction Control Thrusters were assumed to operate on the same propellant as the Auxiliary thrusters.

The Biprop option was selected because of an overall lower subsystem mass due to the higher engine specific impulse. Additionally, the four cylindrical tanks for the propellants were easier to fit into the S/C volume than the larger volume of the single hydrazine tank.

3.6.3.2   REP Stage

The trades to be considered in designing the two major propulsion systems on the S/C (main and RCS) are as follows

- Main REP propulsion system

- ~1000 W Xe Hall or Ion thruster
  - Two for life, one more for spare
  - Direct Drive System
- Up to 600 kg Xe stored in COPV tanks

The possible main EP system options to be considered for this design are
- New Advanced Technology Small Hall Thruster (Figure 3.11)
  - Based on ongoing HiVHAC program at GRC
  - Optimized design to allow up to 2000 s I_{sp} at powers below 1 kW
  - Allows long life needed for mission
- Derated HiVHAC
  - Maximum I_{sp} at 1 kW ~1570 s
  - Performance inadequate for range of REP missions
- Commercial-off-the-shelf (COTS)
  - SPT-70/BPT-600
  - 600 W, ~ 1500 s
  - Limited life/throughput (35 to 50 kg)
- Low power (20 cm) Ion

The Advanced Hall thruster option was initially chosen both for its potential for Direct Drive operation (see power system discussion), and for its superior performance in terms of efficiency (or equivalently, thrust-to-power) at the low power levels characteristic of REP. The 20 cm ion thruster projected performance was inferior to that of the Hall below 1 kW and at 2000 s or less I_{sp}. The commercial Hall thrusters increased system mass and complexity through the increased number of propulsion strings (13 or more) needed to meet lifetime and redundancy requirements.

The possible EP thruster system options, once an EP thruster type has been chosen, to be considered are

- Hall
  - Standard PPU
  - "Direct Drive" from Stirling Alternator

Because of limitations in the ASRG alternator design, the "Direct Drive" option was discarded and a standard PPU option was selected.

Figure 3.11.—Advanced Technology
Small Hall Thruster.

3.6.4 Propulsion and Propellant Management Design and MEL

3.6.4.1 Chemical Cruise Stage

The cruise stage propulsion system has two functions. First, the auxiliary propulsion is used for orbit shaping which are performed with two Aerojet HiPAT (R-4D-derived thruster) bipropellant (MMH/NTO) thrusters operating at 441 N. The orbit shaping maneuvers are performed with a single thruster. The second thruster was operated in the event of a fault with the first engine.

Second, S/C attitude control requirements are performed with a set of twelve Rocketdyne R-53 8.9 N bipropellant thrusters. These thrusters were arranged in four pods of three thrusters that are arranged orthogonally in order to provide all attitude control thrusting.

As mentioned above, the MMH fuel and NTO oxidizer are stored in two cylindrical carbon over-wrapped (COPV) tanks each. It was decided to use two tanks for each liquid in order ease the volumetric conflicts with other elements in the S/C. Each tank is wrapped with tape heaters in order to maintain the required propellant temperatures. The propellant is delivered from the tanks to the thrusters through 3/8-in. lines that are also wrapped with tape heaters to maintain propellant temperatures. While the propellant lines are zero-fault tolerant, each of the isolation and control valves are dual-seat to provide single fault tolerance.

The propellant tanks were pressurized with He gas that is stored at high pressure in a single metallic tank for fuel and oxidizer tanks, respectively. The He management subsystem is single-fault tolerant with dual-seat isolation valves and redundant pressure regulators. However, the He lines are zero-fault tolerant. Figure 3.12 shows a schematic of the propulsion subsystem.

Table 3.13 lists the MEL for the propulsion system in the chemical cruise stage. All growth allowances follow the AIAA WGA schedule in Section 2.4.

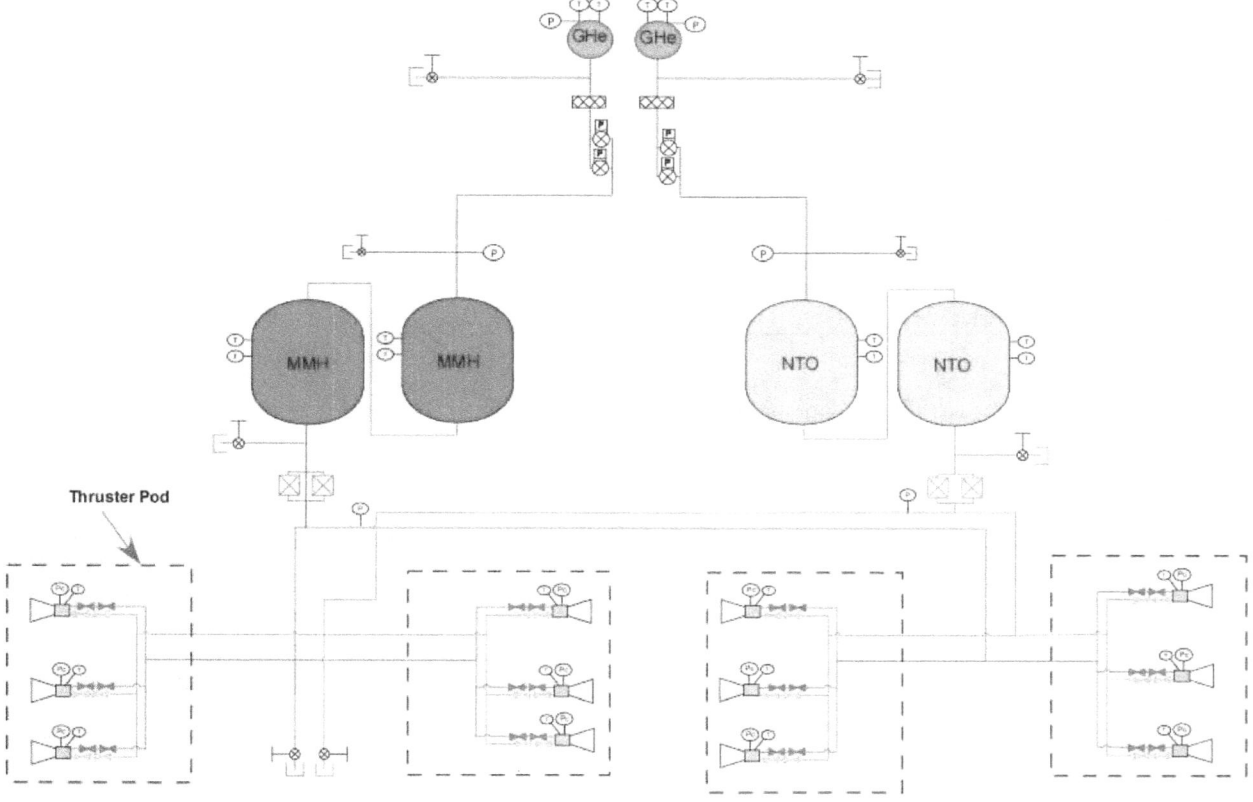

2 lbf Thrusters
Figure 3.12.—RCS Configuration for Cruise Stage.

Table 3.14 lists the MEL for the propellant in the chemical cruise stage. Note that the propellant is modeled as an RCS system. All growth allowances follow the AIAA WGA schedule in Section 2.4.

TABLE 3.13.—PROPULSION AND PROPELLANT MANAGEMENT SYSTEM MEL CHEMICAL CRUISE STAGE

WBS no.	Description Generic Chemical Stage	Qty	Unit mass (kg)	CBE mass (kg)	Growth (%)	Growth (kg)	Total mass (kg)
03	Cruise Deck/Chemical Stage	-	-------	896.63	-------	57.23	953.86
03.6	Propulsion	-	-------	74.38	30.00	22.31	96.69
03.6.1	Primary chemical system hardware	-	-------	13.55	30.00	4.06	17.61
03.6.1.a	Main engine	-	------	13.55	30.00	4.06	17.61
03.6.1.a.a	Main engine	2	6.12	12.25	30.00	3.67	15.92
03.6.1.a.b	Main engine gimbal	2	0.65	1.30	30.00	0.39	1.69
03.6.1.a.c	Miscellaneous no. 1	0	0.00	0.00	0.00	0.00	0.00
03.6.1.a.d	Miscellaneous no. 2	0	0.00	0.00	0.00	0.00	0.00
03.6.2	Propellant management (chemical)	-	-------	50.47	30.00	15.14	65.61
03.6.2.a	Main engine propellant management	-	-------	3.34	30.00	1.00	4.34
03.6.2.a.a	Fuel tanks	0	0.00	0.00	0.00	0.00	0.00
03.6.2.a.b	Fuel lines	0	0.00	0.00	0.00	0.00	0.00
03.6.2.a.c	Oxidizer tanks	0	0.00	0.00	0.00	0.00	0.00
03.6.2.a.d	Oxidizer lines	0	0.00	0.00	0.00	0.00	0.00
03.6.2.a.e	Pressurization system—tanks, panels, lines	0	0.00	0.00	0.00	0.00	0.00
03.6.2.a.f	Feed system—regulators, valves, etc	1	3.34	3.34	30.00	1.00	4.34
03.6.2.a.g	Miscellaneous no. 1	0	0.00	0.00	0.00	0.00	0.00
03.6.2.a.h	Miscellaneous no. 2	0	0.00	0.00	0.00	0.00	0.00
03.6.2.b	RCS propellant management	-	-------	47.13	30.00	14.14	61.27
03.6.2.b.a	Fuel tanks	1	30.54	30.54	30.00	9.16	39.71
03.6.2.b.b	Fuel lines	0	0.00	0.00	0.00	0.00	0.00
03.6.2.b.c	Pressurization System—tanks, panels, lines	1	10.01	10.01	30.00	3.00	13.01
03.6.2.b.d	Feed System—regulators, valves, etc	1	6.58	6.58	30.00	1.97	8.55
03.6.2.b.e	Miscellaneous no. 1	0	0.00	0.00	0.00	0.00	0.00
03.6.2.b.f	Miscellaneous no. 2	0	0.00	0.00	0.00	0.00	0.00
03.6.3	RCS hardware	-	-------	10.36	30.00	3.11	13.47
03.6.3.a	RCS engines	4	2.59	10.36	30.00	3.11	13.47
03.6.3.b	RCS thruster subassembly	0	0.00	0.00	0.00	0.00	0.00

TABLE 3.14.—PROPELLANT MEL FOR CHEMICAL CRUISE STAGE

WBS no.	Description Generic Chemical Stage	Qty	Unit mass (kg)	CBE mass (kg)	Growth (%)	Growth (kg)	Total mass (kg)
03	Cruise Deck/Chemical Stage	-	---------	896.63	6.38	57.23	953.86
03.7	Propellant (chemical)	-	---------	642.63	0.00	0.00	642.63
03.7.1	RCS propellant	-	---------	642.63	0.00	0.00	642.63
03.7.1.a	Fuel	-	---------	241.38	0.00	0.00	241.38
03.7.1.a.a	Fuel usable	1	235.49	235.49	0.00	0.00	235.49
03.7.1.a.b	Fuel boiloff	0	0.00	0.00	0.00	0.00	0.00
03.7.1.a.c	Fuel residuals (unused)	1	5.89	5.89	0.00	0.00	5.89
03.7.1.b	Oxidizer	-	---------	398.27	0.00	0.00	398.27
03.7.1.b.a	Oxidizer usable	1	388.56	388.56	0.00	0.00	388.56
03.7.1.b.b	Oxidizer boiloff	0	0.00	0.00	0.00	0.00	0.00
03.7.1.b.c	Oxidizer residuals (unused)	1	9.71	9.71	0.00	0.00	9.71
03.7.1.c	RCS pressurant	1	2.99	2.99	0.00	0.00	2.99

3.6.4.2   REP Powered Main EP System (Xe) for Spiral at Saturn System

The main EP system is comprised of:

- Four extended life, High I_{sp} Hall Thrusters (one operating)
    - One active 1000 W Long Life Hall engine
        - Two extra thrusters for life and one cold spare
    - 30,000 hr life, 300 to 600 V
- Dedicated PPU for each thruster; no cross-strap
- Dedicated gimbal for each thruster
    - Two-axis range of motion: $\pm XX°$, $\pm ZZ°$
- One spherical COPV high pressure (2800 psi) Xe tank
- Propellant distribution system: Single string (zero-fault tolerant) Propellant Management System (PMS) to each thruster from balanced tank feed
- Thermal control system for propellant management subsystem
    - Tape heaters and insulation used on Xe tank and feedlines
- Total propellant
    - 540 kg used
    - 8.6 percent residual + margin unusable Xe

The Advanced Hall thruster assumed to be 50 percent heavier than SPT-70 thruster and have 30,000-hr life capabilities.

Figure 3.13 is a schematic of the EP system and propellant management tankage, etc. The main electric propulsion subsystem is comprised of: four HiVHAC Hall Thrusters—three operating, one spare, Gimbals on each thruster for thrust vector control, one carbon-overwrapped (COPV) titanium-lined high-pressure cylindrical storage tank for the Xe propellant (nominal), Xe distribution system based on newly developed pressure and flow control units and four PPU for delivering power to each ion thruster.

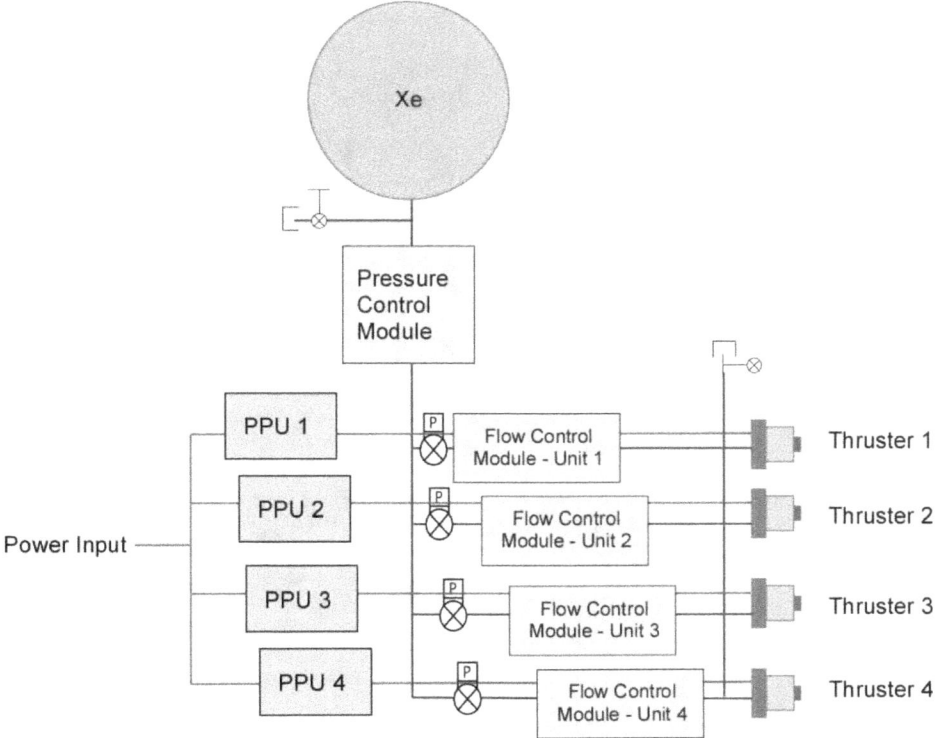

Figure 3.13.—REP S/C main electrical propulsion subsystem configuration.

The REP S/C chemical propulsion subsystem had two functions: 1) reaction/attitude control, 2) orbit capture at destination. Hydrazine has been baselined as the propellant for this thruster system, which is comprised of

- Twelve MR-103 thrusters positioned over the body of the S/C
  - Aerojet hydrazine thrusters, TRL9; JPL "Minimum Impulse Thruster (MIT)" engine
  - Thruster operates at 1 N, 227 s I_{sp}
- Two MR-104A/C thrusters positioned side-by-side on the vehicle side opposite that of the main communication array
  - Aerojet hydrazine thruster, TRL 9
  - Thruster operates at 441 N, 239 s I_{sp}
  - Single thruster is used, with second engine as spare
- Hydrazine propellant stored in two cylindrical COPV tanks
  - Similar to previous NH system
- Propellant distribution system similar to Cruise Stage RCS
- Propellant tanks included blow down pressurization to drive thrusters
- Thermal protection approach
  - Propellant tanks and feedlines are covered with tape heaters and insulation to maintain propellant as liquid
- Total propellant
  - Useable 205 kg
  - 8 percent margin added

Figure 3.14 shows the configuration of the chemical propulsion subsystem.

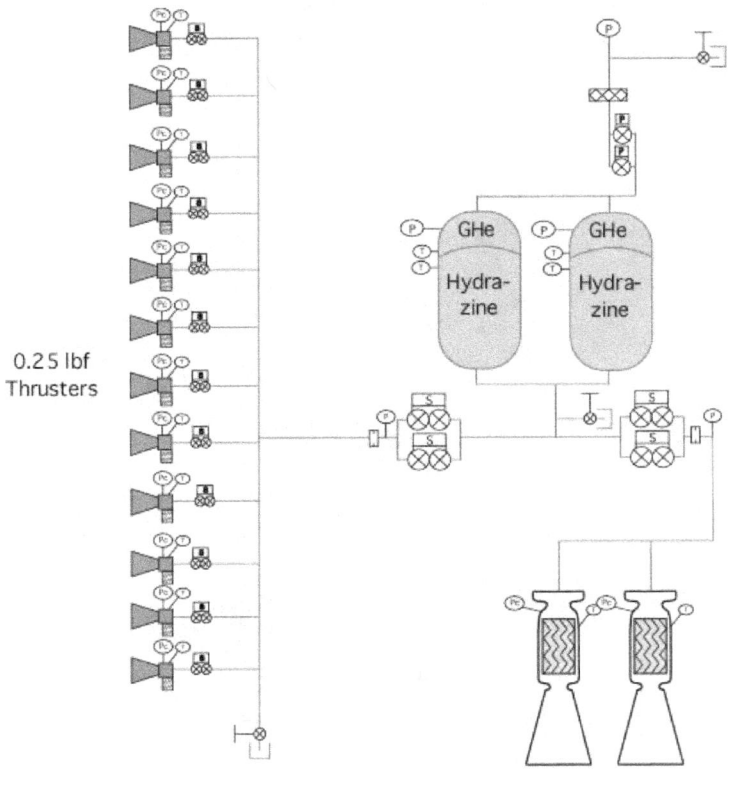

0.25 lbf
Thrusters

100 lbf
Thrusters
Figure 3.14.—REP S/C chemical propulsion subsystem.

Table 3.15 lists the propulsion system hardware MEL for the REP stage at Saturn. All growth allowances follow the AIAA WGA schedule in Section 2.4 and do not contain the additional 8.8 percent carried at the system level.

TABLE 3.15.—PROPULSION SYSTEM MEL REP S/C INSIDE AEROSHELL

WBS no.	Description REP Flagship Mission Aeroshell (November 7, 2007)	Qty	Unit mass (kg)	CBE mass (kg)	Growth (%)	Growth (kg)	Total mass (kg)
01	REP S/C (Payload and Stage)	---	-------	1784.15	10.90	194.52	1978.67
01.2.6	Propulsion	---	-------	145.64	28.22	41.10	186.74
01.2.6.a	Primary EP system	---	-------	6.75	12.00	0.81	7.56
01.2.6.a.a	Primary EP thrusters	3	2.25	6.75	12.00	0.81	7.56
01.2.6.a.b	EPS power processing and control	0	0.00	0.00	0.00	0.00	0.00
01.2.6.a.c	EPS structure	---	-------	0.00	0.00	0.00	0.00
01.2.6.a.c.a	EP thruster pod	0	0.00	0.00	0.00	0.00	0.00
01.2.6.a.c.b	EP thruster boom	0	0.00	0.00	0.00	0.00	0.00
01.2.6.a.c.c	Miscellaneous no. 1	0	0.00	0.00	0.00	0.00	0.00
01.2.6.a.d	EPS thermal control subsystem	---	-------	0.00	0.00	0.00	0.00
01.2.6.a.d.a	EPS MLI	0	0.00	0.00	0.00	0.00	0.00
01.2.6.a.d.b	EPS heaters and sensors	0	0.00	0.00	0.00	0.00	0.00
01.2.6.a.d.c	Miscellaneous no. 1	0	0.00	0.00	0.00	0.00	0.00
01.2.6.b	Propellant management	---	-------	57.41	31.36	18.00	75.42
01.2.6.b.a	Xe propellant tank(s)	1	44.61	44.61	30.00	13.38	58.00
01.2.6.b.b	High pressure feed system	1	8.90	8.90	30.00	2.67	11.57
01.2.6.b.c	Low pressure feed system	0	0.00	0.00	0.00	0.00	0.00
01.2.6.b.d	Residual Xe propellant (nondeterministic)	0	0.00	0.00	0.00	0.00	0.00
01.2.6.b.e	Temperature sensors	1	3.90	3.90	50.00	1.95	5.85
01.2.6.b.f	Propulsion tank heaters	0	0.00	0.00	0.00	0.00	0.00
01.2.6.b.g	Propulsion line heaters	0	0.00	0.00	0.00	0.00	0.00
01.2.6.b.h	Miscellaneous no. 1	0	0.00	0.00	0.00	0.00	0.00
01.2.6.b.i	Miscellaneous no. 2	0	0.00	0.00	0.00	0.00	0.00
01.2.6.b.j	Miscellaneous no. 3	0	0.00	0.00	0.00	0.00	0.00
01.2.6.b.k	Miscellaneous no. 4	0	0.00	0.00	0.00	0.00	0.00
01.2.6.c	PPU	---	-------	12.00	12.00	1.44	13.44
01.2.6.c.a	PPU mass	3	4.00	12.00	12.00	1.44	13.44
01.2.6.c.b	Cabling	0	0.00	0.00	0.00	0.00	0.00
01.2.6.c.c	Miscellaneous no. 1	0	0.00	0.00	0.00	0.00	0.00
01.2.6.c.d	Miscellaneous no. 2	0	0.00	0.00	0.00	0.00	0.00
01.2.6.c.e	Miscellaneous no. 3	0	0.00	0.00	0.00	0.00	0.00
01.2.6.d	RCS	---	-------	69.48	30.00	20.84	90.32
01.2.6.d.a	RCS tank subassembly	1	30.54	30.54	30.00	9.16	39.71
01.2.6.d.b	RCS propellant management subassembly	1	24.45	24.45	30.00	7.34	31.79
01.2.6.d.c	RCS thruster subassembly	12	1.21	14.48	30.00	4.34	18.82

Table 3.16 lists the propellant used in this mission. Note, the margins and residuals are called out as separate line items in this mass listing, and no additional WGS is necessary on the propellants. Figure 3.15 shows the power and propulsion deck of the REP S/C. The ASRGs are mounted to the bus structure via trusses, at a 45° angle between each other radially around the main bus.

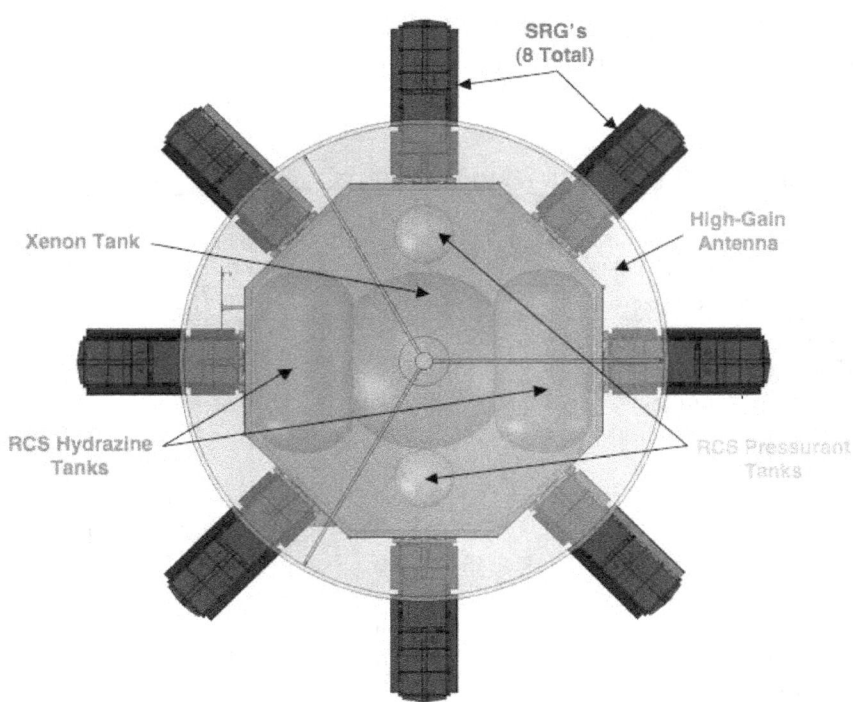

Figure 3.15—Flagship mission REP stage propulsion deck.

TABLE 3.16.—EP PROPELLANT MEL FOR REP STAGE

WBS no.	Description REP Flagship Mission Aeroshell (November 7, 2007)	Qty	Unit mass (kg)	CBE mass (kg)	Growth (%)	Growth (kg)	Total mass (kg)
01	REP S/C (Payload and Stage)	-	--------	1784.15	10.90	194.52	1978.67
01.1	Science Payload	-	--------	70.00	30.00	21.00	91.00
01.2	REP Bus	-	--------	1714.15	10.12	173.52	1887.67
01.2.7	Propellant	-	--------	847.51	0.00	0.00	847.51
01.2.7.a	Primary EP propellant	-	--------	636.57	0.00	0.00	636.57
01.2.7.a.a	Primary EP propellant used	1	586.16	586.16	0.00	0.00	586.16
01.2.7.a.b	Primary EP propellant residuals (unused)	1	21.10	21.10	0.00	0.00	21.10
01.2.7.a.c	Primary EP propellant performance margin (unused)	1	29.31	29.31	0.00	0.00	29.31
01.2.7.b	RCS propellant	-	--------	210.18	0.00	0.00	210.18
01.2.7.b.a	RCS used	1	205.05	205.05	0.00	0.00	205.05
01.2.7.b.b	RCS residuals	1	5.13	5.13	0.00	0.00	5.13
01.2.7.c	Pressurant	1	0.77	0.77	0.00	0.00	0.77

3.6.5   Propulsion and Propellant Management Analytical Methods

Hall thruster and PPU performance and masses were based on published or in-house calculations by the GRC's In Space Propulsion branch. Thruster performance over a range of specific impulse was examined as a series of custom designs, rather than a single thruster design capable of throttling over a range of I_{sp} (Figure 3.16). Thruster mass was assumed to be 50 percent greater than a commercial thruster (SPT-70) operating at a similar power level. PPU performance and mass were based on a single module of a PPU unit under development and test at GRC.

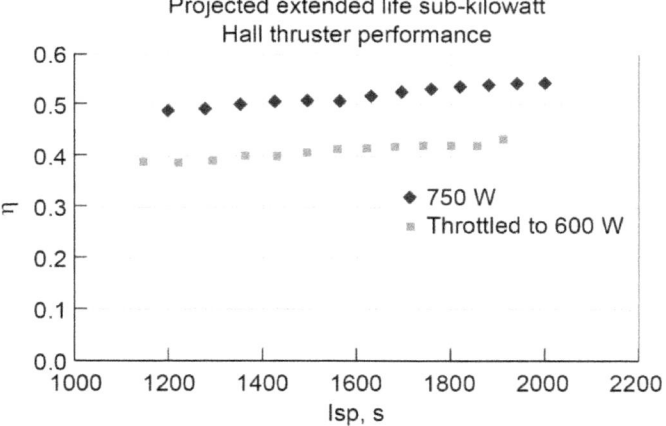

Figure 3.16.—Flagship mission EP thruster performance assumption.

3.6.6 Propulsion and Propellant Management Risk Inputs

There were three risk issues that were identified for detailed risk analysis. These issues are

- Failure of the electric thruster to start/operate.
- Impingement of RCS thruster plumes on sensitive S/C surfaces
- Freezing of chemical propellant in lines or tanks

3.6.7 Propulsion and Propellant Management Recommendation

Future trades to on the main propulsion system are as follows

- Derated HiVHAC (mass, trip time penalties)
- Low power ion thruster (8 cm)

Further trades on the secondary propulsion system to reduce mass are

- Utilize primary propulsion for some maneuvering, and modeling the trade between attitude control using Wheels versus propulsion.

3.7 Thermal Control

3.7.1 Thermal Requirements

The thermal requirements are dominated by the need to place the ASRG system inside of the aeroshell during transit and aerocapture itself. A substantial amount of waste heat (125 Wth) needs to be removed from the ASRGs while inside of the aeroshell. Once the Aeroshell is jettisoned, the ASRGs will see deep space and Titan or Enceladus' surface and no longer have thermal issues.

The thermal requirements after the aerocapture were to provide a means of cooling the S/C during operation as well as provide heat to vital components and systems to maintain a minimum temperature throughout the mission.

The maximum heat load to be rejected by the thermal system was 125 W from the electronics, and the desired operating temperature for the electronics and propellant was 300 K. The ASRGs have dedicated built in thermal control systems and therefore were not part of the S/C thermal system.

3.7.2 Thermal Assumptions

The thermal subsystem approach was based on that used in the COMPASS SEP Enceladus design (CD–2007–08 (Ref. 11)).

The assumptions utilized in the analysis and sizing of the thermal system were based on the operational environment. It was assumed that the worst case operational conditions would be in near Earth space. The following assumptions were utilized to size the thermal system.

- The view factors for the radiator to the Earth, lunar surface and SRG radiators were assumed to be 0.1, 0.25, and 0.1 respectively.
- The maximum angle of the radiator to the Sun was 15°.
- The radiator temperature was 320 K.

During cruise and prior to aerocapture a radiator system mounted on the REP stage deck will be used to reject the heat from all the internal systems including the ASRGs. After aerocapture and jettison of the SEP stage a separate thermal control system will be used to reject heat from the S/C components. This second radiator system is housed inside the aeroshell and is not operational until after the aerocapture.

3.7.3 Thermal Design and MEL

3.7.3.1 Cruise and Aerocapture Phase

During the cruise phase the ASRGs and avionics are inside of the aeroshell. In order to cool them sufficiently a pumped loop radiator system is used to connect the ASRGs to a radiator on the cruise deck which is to be jettisoned before aerocapture. A separate loop cools the lower temperature avionics on the REP S/C using the cruise deck radiator.

Since the cruise deck must be jettisoned for the aeroshell to work properly, no external cooling will be available during the 10 to 40 min trip through Titan's upper atmosphere. This transit will also incur additional heat radiated from the ionized atmospheric nitrogen. It is assumed that the aero backshell will protect the REP S/C sufficiently. The heat of the operating ASRGs and avionics is assumed to be cooled sufficiently during this short time by boiling off the radiator coolant still in the lines.

Table 3.17 lists the items in the Thermal system MEL for the COMPASS REP S/C design. All growth allowances follow the AIAA WGA schedule in Section 2.4 and do not contain the additional 8.8 percent carried at the system level.

TABLE 3.17.—THERMAL MEL FOR CHEMICAL CRUISE STAGE

WBS no.	Description Generic Chemical Stage	Qty	Unit mass (kg)	CBE mass (kg)	Growth (%)	Growth (kg)	Total mass (kg)
03	Cruise Deck/Chemical Stage	-	------	896.63	6.38	57.23	953.86
03.5	Thermal control (non-propellant)	-	------	48.93	15.00	7.34	56.27
03.5.3	Semi-passive thermal control (cruise deck)	-	------	48.93	15.00	7.34	56.27
03.5.3.a	Louvers	0	0.00	0.00	15.00	0.00	0.00
03.5.3.b	Thermal switches	0	0.00	0.00	15.00	0.00	0.00
03.5.3.c	RTG radiator	1	37.70	37.70	15.00	5.66	43.36
03.5.3.d	RTG coolant loop	1	11.23	11.23	15.00	1.68	12.91
03.5.3.e	RTG cold plates	0	3.46	0.00	15.00	0.00	0.00

Table 3.18 lists the items in the Thermal system MEL for the COMPASS REP S/C design. All growth allowances follow the AIAA WGA schedule in Section 2.4 and do not contain the additional 8.8 percent carried at the system level.

TABLE 3.18.—THERMAL MEL FOR REP/AEROCAPTURE

WBS no.	Description REP Flagship Mission Aeroshell (November 7, 2007)	Qty	Unit mass (kg)	CBE mass (kg)	Growth (%)	Growth (kg)	Total mass (kg)
01	REP S/C (Payload and Stage)	---	------	1784.15	10.90	194.52	1978.67
01.2.5	Thermal control (nonpropellant)	---	------	61.57	15.00	9.23	70.80
01.2.5.a	Active thermal control	---	------	16.90	15.00	2.54	19.44
01.2.5.a.a	Heaters	15	1.00	15.00	15.00	2.25	17.25
01.2.5.a.b	Thermal control/heaters circuit	2	0.20	0.40	15.00	0.06	0.46
01.2.5.a.c	Data acquisition	1	1.00	1.00	15.00	0.15	1.15
01.2.5.a.d	Thermocouples	50	0.01	0.50	15.00	0.08	0.58
01.2.5.a.e	Miscellaneous no. 1	0	0.00	0.00	15.00	0.00	0.00
01.2.5.a.f	Miscellaneous no. 2	0	0.00	0.00	15.00	0.00	0.00
01.2.5.b	Passive thermal control	---	------	41.17	15.00	6.18	47.34
01.2.5.b.a	Heat sinks	4	3.46	13.85	15.00	2.08	15.93
01.2.5.b.b	Heat pipes	1	1.37	1.37	15.00	0.21	1.58
01.2.5.b.c	Radiators	1	4.69	4.69	15.00	0.70	5.39
01.2.5.b.d	MLI	1	14.51	14.51	15.00	2.18	16.69
01.2.5.b.e	Temperature sensors	25	0.01	0.25	15.00	0.04	0.29
01.2.5.b.f	Phase change devices	0	0.00	0.00	15.00	0.00	0.00
01.2.5.b.g	Thermal coatings/paint	1	3.60	3.60	15.00	0.54	4.14
01.2.5.b.h	Micrometeor shielding	0	0.00	0.00	15.00	0.00	0.00
01.2.5.b.i	S/C RTG MLI	1	0.00	0.00	15.00	0.00	0.00
01.2.5.b.j	S/C engine MLI	1	2.90	2.90	15.00	0.43	3.33
01.2.5.c	Semi-passive thermal control	---	------	3.50	15.00	0.52	4.02
01.2.5.c.a	Louvers	1	2.70	2.70	15.00	0.40	3.10
01.2.5.c.b	Thermal switches	4	0.20	0.80	15.00	0.12	0.92

3.7.4 Thermal Trades

Prior to aerocapture, the thermal system utilizes a pump loop cooling system connected to a radiator on the REP stage deck. This cooling system removes excess heat from all of the ASRGs and the S/C components.

After the aerocapture, the thermal system is used to remove excess heat from the electronics and other components of the system as well as provide heating to thermally sensitive components during periods of inactivity.

Excess heat is collected from a series of Al cold plates located throughout the interior of the S/C. These cold plates have heat pipes integrated into them. The heat pipes transfer heat from the cold plates to the radiator, which radiates the excess heat to space. The portions of the heat pipes that extend from the S/C body and are integrated to the radiator are protected with a micro meteor shield. The radiator has exterior louvers on it to provide some control over its heat transfer capability.

The radiator was sized with approximately 50 percent margin in its heat rejection area. This added margin insures against unforeseen heat loads, degradation of the radiator and increased view factor toward the sun or other thermally hot body not accounted for in the analysis.

To provide internal heating for the electronics and propulsion systems a series of electric heaters are utilized. These heaters are controlled by an electronics controller, which reads a series of thermocouples through a data acquisition system.

MLI is also utilized on the S/C, and propellant system to regulate and maintain the desired temperatures.

3.7.5　Thermal Analytical Methods

Objective

To provide spreadsheet based models capable of estimating the mass and power requirements of the various thermal systems. The models were produced based on a first principles analysis. They were structured to be easily adapted to various missions.

The analysis performed to size the thermal system is based on first principle heat transfer from the S/C to the surroundings. This analysis takes into account the design and layout of the thermal system and the thermal environment to which heat is being rejected or the vehicle is being insulated from. For more detailed information on the thermal analysis, a summary white paper entitled "Spacecraft Preliminary Thermal System Sizing for Trade Study Analysis" was produced. This paper is presently under publication as a NASA Contractor Report.

The thermal for the Aerobraking portion of the mission was based on the Mars Science Laboratory (MSL). Radiator pump loop cooling was used on the REP S/C while inside the aeroshell.

3.7.5.1　Thermal Environmental Models

Calculations of solar intensity are based on S/C location. A standard solar intensity model was used to determine the solar flux on the S/C throughout the mission profile. This model is detailed in the report titled "Spacecraft Preliminary Thermal System Sizing for Trade Study Analysis."

3.7.5.2　Thermal Systems Modeled

Figure 3.17 shows a conceptual design of the Cruise Deck (stage) thermal system. This system consists of a pump loop coolant system and radiator panel. The coolant system rejects heat from the ASRGs and the S/C internal components during the cruise phase of the mission prior to aerocapture.

After the aerocapture maneuver is completed a heat pipe based thermal control system is utilized to maintain the S/C components at their desired temperatures. The main components from this system are illustrated in Figure 3.18.

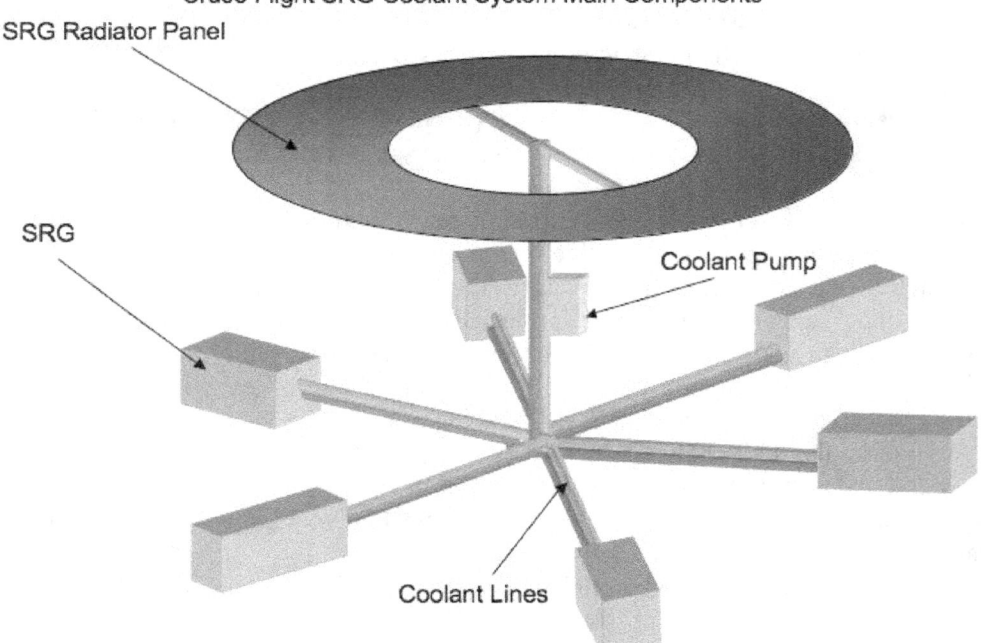

Figure 3.17.—Cruise Deck thermal protection system.

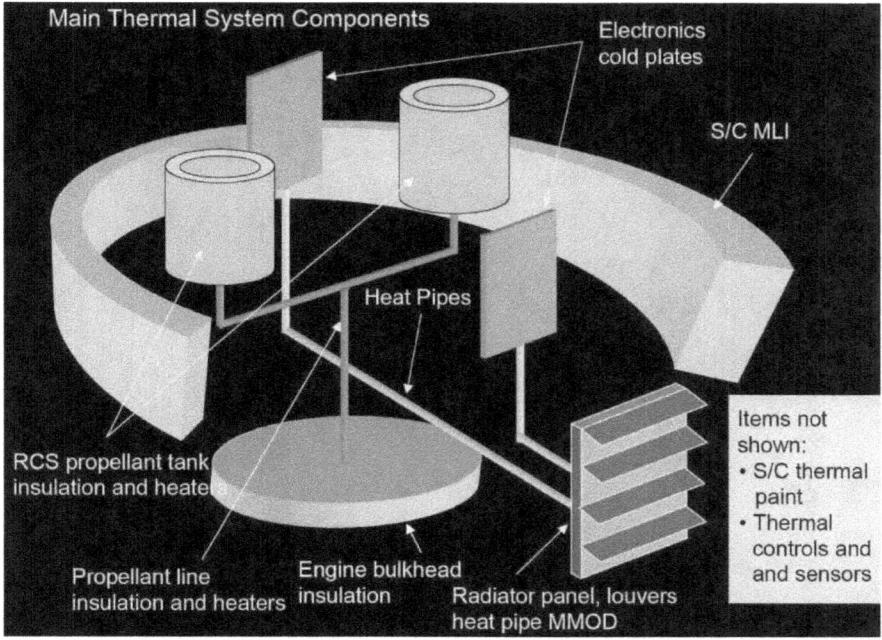

Figure 3.18—Illustration of the main components of the S/C thermal control system.

3.7.5.2.1   Micrometeor Shielding on Radiator

No specific micrometeor shielding was used on the S/C. The aeroshell structure was deemed sufficient to protect the vehicle during the near earth and cruse phases. The radiator coolant lines that were outside of the aeroshell were the only components that had dedicated micrometeor shielding.

3.7.5.2.2   Radiator Panel Modeling

- The radiator panel area has been modeled along with a rough estimate of its mass.
- The model was based on a first principles analysis of the area needed to reject the identified heat load to space. From the area a series of scaling equations were used to determine the mass of the radiator.
- Worst-case thermal environment was used to size the radiator.

TABLE 3.19.—RADIATOR THERMAL ASSUMPTIONS

Variable	Value
Radiator solar absorptivity	0.14
Radiator emissivity	0.84
Radiator Sun angle	90°
Radiator operating temperature	320 K
S/C radiation dissipation power	250 W
SRG radiator thermal dissipation	2800 W

3.7.5.2.3   Thermal Control of Propellant Lines and Tanks

- Power requirements and mass have been modeled. This modeling included propellant tank MLI and heaters and propellant line insulation and heaters.
- Worst-case thermal environment was used to calculate the heat loss.
- The model was based on a first principles analysis of the radiative heat transfer from the tanks and propellant lines to space. The heat loss through the insulation set the power requirement for the tank and line heaters. See Figure 3.19 for details of the insulation mass versus number of insulation layers and the subsequent heat loss for that number of layers. Figure 3.19 is used to size the mass of the MLI used in the thermal protection system.

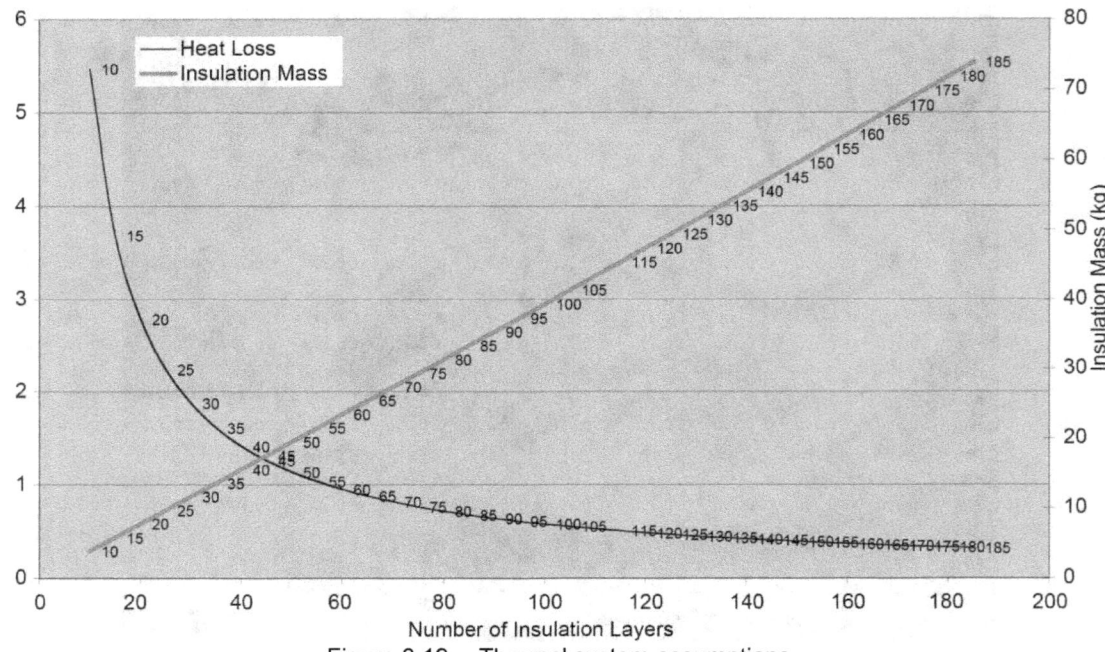
Figure 3.19.—Thermal system assumptions.

Table 3.20 lists the environmental and engineering assumptions used in the tank insulation calculations.

TABLE 3.20.—THERMAL ASSUMPTIONS FOR TANK INSULATION CALCULATIONS

Variable	Value
Tank surface emissivity (ε_t)	0.1
MLI emissivity (ε_i)	0.07
MLI material	Al
MLI material density (ρ_i)	2,770 kg/m^3
Internal tank temperature (T_i)	300 K
MLI layer thickness (t_i)	0.025 mm
Number of insulation layers (n_i)	10
MLI layer spacing (d_i)	1.0 mm
Tank immersion heater mass and power level	1.02 kg at up to 1,000 W
S/C inner wall surface emissivity	0.98
S/C outer wall surface emissivity	0.93
Line foam insulation conductivity	0.0027 W/m K
Line foam insulation emissivity	0.07
Propellant line heater specific mass and power	0.143 kg/m at up to 39 W/m
Line foam insulation density	56 kg/m^3

3.7.5.2.4   S/C Insulation

The mass of the S/C MLI on the engine bulkhead MLI was modeled to determine the mass of the insulation and heat loss. The model was based on a first principles analysis of the heat transfer from the S/C through the insulation to space. The near earth thermal environment was used to size the insulation. Table 3.21 lists the thermal environment assumptions used in sizing the S/C insulation.

TABLE 3.21.—THERMAL ENVIRONMENT
ASSUMPTIONS FOR S/C INSULATION

Variable	Value
S/C MLI material .... Al	
S/C MLI material density (ρ_{isc})...2770 kg/m^3	
MLI layer thickness (t_i) .... 0.025 mm	
Number of insulation layer(n_i).. 25	
MLI layer spacing (d_i) .... 1.0 mm	
S/C radius (r_{sc}).. 1.145 m	

3.7.5.2.5   Avionics and Power Management and Distribution (PMAD) Cooling

Mass estimates for the Active Thermal Cooling System (ATCS) have been completed. The components of the system included cold plates and heat pipes. The model was based on a first principles analysis of the area needed to reject the identified heat load to space. From the sizing a series of scaling equations were used to determine the mass of the various system components (Table 3.22).

TABLE 3.22.—THERMAL ENVIRONMENT ASSUMPTIONS
FOR AVIONICS AND PMAD COOLING

Variable	Value
Cooling plate and lines material .... Al	
Cooling plate and lines material density......................................2770 kg/m^3	
Number of cooling plates 2	
Cooling plate lengths.. 0.5 m	
Cooling plate widths.. 0.5 m	
Cooling plate thickness... 5 mm	
Heat pipe specific mass .... 0.15 kg/m	

3.7.5.3   Thermal Assumptions

- Deep Space operation
- Radiator always sees deep space with a small (0.05) view factor to the Sun
- The ASRGs produced 2800 W waste heat for rejection by the radiators. The communications/avionics/EP system produced a maximum of 250 W of waste heat for rejection by the radiators.

3.7.6   Thermal Risk Inputs

The risks associated with the thermal system are based mainly on the failure of a component of multiple components of the system. The majority of the system operation is passive and therefore has a fairly high reliability. Some of the major failure mechanisms are listed below.

- Pumped coolant loop failure. This can occur due to a leak in the coolant line or a pump failure. This type of failure would cause a loss of the mission. Heat pipe failure. This can be due to cracking due to thermal stresses, micrometeor impact or design defect. This likelihood of this type of failure is low. The impact of this failure would be a loss of all or a portion of the S/Cs capability.
- Heater system failure. This would most likely be due to wire breakage or a controller failure. The likelihood of this type of failure is low. The impact of this failure would be a loss of certain components or propulsion capability once the vehicle is exposed to an extended period of cold.

- Radiator louver failure. The thermal controller on the system can fail due to an electronics failure or power failure. Subsequently this will cause a failure of the radiator louvers reducing the effectiveness of the radiator and limiting control of the thermal system. The louvers can also experience mechanical failure causing them to be held in a fixed position or limiting their range of motion.

3.7.7 Thermal Recommendation

To improve the reliability of the system and compensated for the identified failure risks the following system design changes can be made.

- A redundant coolant loop and coolant pumps can be utilized to minimize the risk of a coolant system failure during the cruse portion of the mission.
- Redundant heat pips can be utilized for each cold plate. The heat pipes can be individually run to the radiator to provide independent cooling paths. The radiator can be separated into two independent units providing additional redundancy.
- Redundant heating system controllers can be utilized. The heaters can be wired individually so that a single heater failure does not bring down any additional heaters. Additional insulation can be added to the S/C to insure that the interior components do not drop below their desired minimum temperature based on a know shadow period of operation.
- The radiator louvers can be designed to fail opened to a specified angle. This will enable the radiator to continue to operate, although not optimally, for the remainder of the mission.

4.0 Cost

4.1 Costing: Flagship Configuration

4.1.1 Costing Assumptions

S/C costs reflect mean estimate (approximately 60 percent confidence level). Quantitative risk analysis on S/C cost based on potential mass growth and CER uncertainty. The ASRG is assumed to be flight ready by its own development project. The S/C fee assumed at 10 percent and is not applied to science instruments (assumed to be furnished equipment). NASA project office/technical oversight are based on 5 percent of all other costs. Phase A based on 5 percent of S/C costs. Mission ops cost assume 9-yr cruise to Titan, 1 yr of science at Titan, 5-yr cruise to Enceladus, and 1 yr of science at Enceladus. Reserves of 30 percent have not been applied to Launch Services or Mission Ops costs. Costs for aerocapture technology development and demonstration are not included.

4.1.2 Cost Estimates

Table 4.1 lists the life cycle cost estimates for this REP Flagship mission. Table 4.2 lists the S/C Estimate Details.

TABLE 4.1.—LIFE CYCLE COST—FY08 ($M)

NASA Project Office/Technical Oversight	71
Phase A	39
S/C with Science Instruments	715
S/C Prime Contractor Fee	65
Launch Services	250
Mission Operations	350
Contingency (30%)	267
Life Cycle Cost	1,757

TABLE 4.2.—S/C ESTIMATE DETAILS—FY08 $M

WBS element	Element name	DDT&E total	Flight hardware	S/C total
REP S/C (Payload and Stage)				
0.1.1	Science Payload	34.4	28.1	62.4
01.2.1	Attitude Determination and Control	10.0	11.1	21.1
01.2.1.a.a	Sun Sensors	1.7	4.0	5.8
01.2.1.a.c	Star Trackers	1.3	2.5	3.8
01.2.1.a.d	IMU	7.0	4.5	11.6
01.2.2	C&DH	30.8	5.1	35.9
01.2.2.a.a	Flight Computer	3.7	4.0	7.6
01.2.2.a.c	Data Interface Unit	0.4	0.5	0.9
01.2.2.a.e	Operations Recorder	0.1	0.1	0.3
01.2.2.a.f	Command and Control Harness (data)	5.5	0.5	6.0
	Flight Software/Firmware	21.1		21.1
01.2.3	Communications and Tracking	14.1	6.8	21.0
01.2.3.a.a	Transmitter/Receiver	4.7	1.8	6.5
01.2.3.a.b	Power Amp	2.1	1.8	3.9
01.2.3.a.d	Antenna	3.5	2.3	5.8
01.2.3.a.h	Cabling	1.0	0.3	1.3
01.2.3.c.a	Coaxial Cable	2.8	0.6	3.5
01.2.4	Electrical Power Subsystem	17.6	189.8	207.5
01.2.4.a.a	RPS Main System	11.4	185.1	196.5
01.2.4.b	Power Management and Distribution	4.3	3.3	7.7
01.2.4.c	Power Cable and Harness Subsystem (C and HS)	1.9	1.4	3.3
01.2.5	Thermal Control (Non-Propellant)	6.8	2.6	9.4
01.2.5.a	Active Thermal Control	0.7	2.1	2.7
01.2.5.b	Passive and Semi-Passive Thermal Control	6.1	0.5	6.6
01.2.6	Propulsion	25.1	14.7	39.8
01.2.6.a.a	Primary EP Thrusters	1.1	1.2	2.3
01.2.6.b.a	Xe propellant tank(s)	4.0	0.9	4.9
01.2.6.b	Balance of propellant Mgt System	4.4	2.5	6.9
01.2.6.c.a	PPU Mass	3.1	2.9	6.0
01.2.6.d.a	RCS Tank Subassembly	2.7	0.6	3.3
01.2.6.d.b	RCS Propellant Management Subassembly	8.0	4.5	12.4
01.2.6.d.c	RCS Thruster Subassembly	1.9	2.0	4.0
01.2.8	Structures and Mechanisms	31.5	13.4	44.9
	Subtotal	170.3	271.6	441.9
Systems integration		89.0	53.0	142.0
Integration, assembly and check out		7.8	10.0	17.8
System test operations		8.4	-------	8.4
Ground support equipment		15.8	-------	15.8
System engineering and integration		28.0	28.6	56.6
Project management		14.4	14.4	28.8
Launch operations and orbital support		14.5	-------	14.5
	Total prime cost	259.3	324.6	583.9
Cruise Deck/Chemical Stage				
03.1	Attitude Determination and Control	2.9	5.0	7.9
03.1.1.a	Sun Sensors	1.6	3.3	5.0
03.1.1.c	Star Trackers	1.2	1.7	2.9

TABLE 4.2.—S/C ESTIMATE DETAILS—FY08 $M

WBS element	Element name	DDT&E total	Flight hardware	S/C total
03.2	C&DH	0.9	0.3	1.1
03.2.1.c	Data Interface Unit	0.4	0.2	0.6
03.2.1.f	Command and Control Harness (data)	0.4	0.1	0.6
03.3	Communications and Tracking	1.1	0.5	1.6
03.3.2	MGA	-------	-------	-------
03.3.2.d	Antenna	0.9	0.4	1.4
03.3.2.e	Cabling	0.2	0.1	0.3
03.4	Electrical Power Subsystem	0.4	0.5	0.9
03.4.3.e	Power Cabling	0.4	0.5	0.9
03.5	Thermal Control (Non-Propellant)	8.8	3.3	12.1
03.5.3.c	RTG Radiator	6.6	1.5	8.1
03.5.3.d	RTG Coolant Loop	2.2	1.8	4.0
03.6	Propulsion	12.3	3.2	15.5
03.6.2.b.a	Fuel Tanks	2.8	0.7	3.5
03.6.2.b.c. d	RCS Propellant Management Subassembly	5.0	1.7	6.7
03.6.3.a	RCS Engines	4.4	0.8	5.3
03.8	Structures and Mechanisms	15.7	4.6	20.4
03.8.1	Structures	9.5	4.3	13.8
03.8.2.e.b	Separation mechanism from LV	2.7	0.1	2.8
03.8.2.e.d	Separation mechanism REP Probe	3.6	0.2	3.7
	Subtotal	42.1	17.4	59.5
Systems integration		26.3	4.6	30.9
Integration, Assembly and Check Out		1.9	0.6	2.5
System Test Operations		2.4	-------	2.4
Ground Support Equipment		4.3	-------	4.3
System Engineering and Integration		8.5	3.1	11.6
Project Management		5.6	0.9	6.5
Launch Operations and Orbital Support		3.5	-------	3.5
	Total prime cost	68.4	22.0	90.4
	Vehicle Integration	19.7	20.8	40.5
	S/C Total	347.4	367.4	714.8

5.0 Trades

Several trades were made including use of REP instead of chemical during cruise from Earth to Saturn, cruise deck equipment (propulsion, navigation, communications), and options for entry probes.

6.0 Challenges, Lessons Learned, Areas for Future Study

A more aggressive mission using an SEP stage based on previous studies may allow for increased delivered mass and potentially science landers at the Saturn moons (Ref. 12).

Appendix A.—Acronyms and Abbreviations

ACS	Attitude Control System		HQ	NASA Headquarters
AD&C	Attitude, Determination and Control		IEM	integrated electronics module
			IMU	Inertial Measuring Unit
AIAA	American Institute for Aeronautics and Astronautics		I_{sp}	Specific Impulse
			JPL	NASA Jet Propulsion Laboratory
Al	aluminum		KSC	NASA Kennedy Space Center
ANSI	American National Standards Institute		Li	lithium
			LSP	Launch Service Program
AO	Announcement of Opportunity		LSTO	Launch Service Task Order
APL	Applied Physics Laboratory		MEL	Master Equipment List
ASRG	Advanced Stirling Radioisotope Generators		MESSENGER	MErcury Surface, Space ENvironment, GEochemistry, and Ranging
ATCS	Active Thermal Cooling System			
BAE	British Aerospace		MGA	Mass Growth Allowance
BOL	beginning of life		MIT	Minimum Impulse Thruster
C&DH	Command and Data Handling		MLI	multilayer insulation
CAD	computer aided design		MMH/NTO	monomethyl hydrazine and nitrogen tetroxide bipropellant system
CBE	current best estimate			
CEV	Crew Exploration Vehicle			
Comm	Communications		MSL	Mars Science Laboratory
COMPASS	COllaborative Modeling and Parametric Assessment of Space Systems		NASA	National Aeronautics and Space Administration
			Nav	navigation
COPV	Composite Overwrapped Pressure Vessel		NEAR	Near Earth Asteroid Rendezvous
			NH	New Horizons
COTS	commercial-off-the-shelf		OTS	off-the-shelf
DSN	Deep Space Network		PEL	Power Equipment List
EGA	Earth Gravity Assist		PMAD	Power Management and Distribution
EIRP	equivalent isotropic radiated power			
ELV	Expendable Launch Vehicle		PMS	Propellant Management System
EOL	end of life		PPU	power processing unit
EP	Electric Propulsion		PSD	Planetary Science Division
FEA	finite element analysis		RBI	Regulator/Bus Protection
FOM	figure of merit		RCS	Reaction Control System
GLIDE	GLobal Integrated Design Environment		REP	Radioisotope Electric Propulsion
			RF	radio frequency
GN&C	Guidance, Navigation and Control		RTG	Radioisotope Thermal Generator
			S/C	spacecraft
GRC	NASA Glenn Research Center		SEAKR	SEAKR Engineering, Inc.
He	helium		SEP	Solar Electric Propulsion
HGA	high gain antenna		TBD	to be discussed

| TWTA | Traveling Wave Tube Amplifier | WGA | Weight Growth Allowance |
| USO | ultra-stable oscillator | Xe | xenon |

Appendix B.—Flagship Design Rendered Drawings

Figure B.1 through Figure B.4 show the REP stage rendered views. Figure B.1 highlights the science instruments. Figure B.2 shows the bottom face of the EP thrusters. Figure B.3 showing face on view of the science instruments side. Figure B.4 also shows the bottom face of the EP thrusters from a slightly different view.

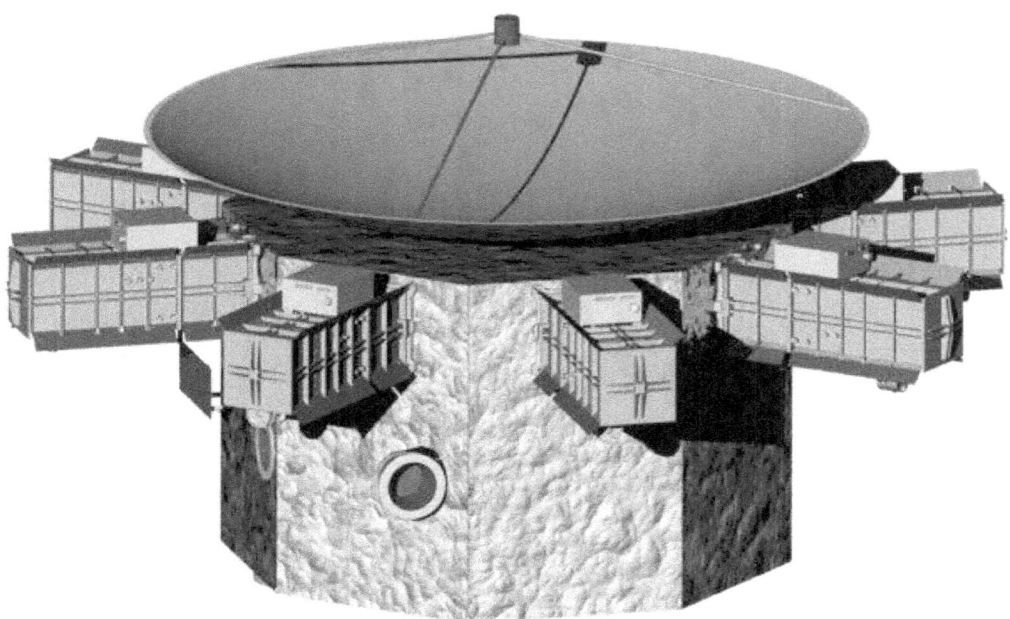

Figure B.1.—REP stage rendered view, showing science instruments.

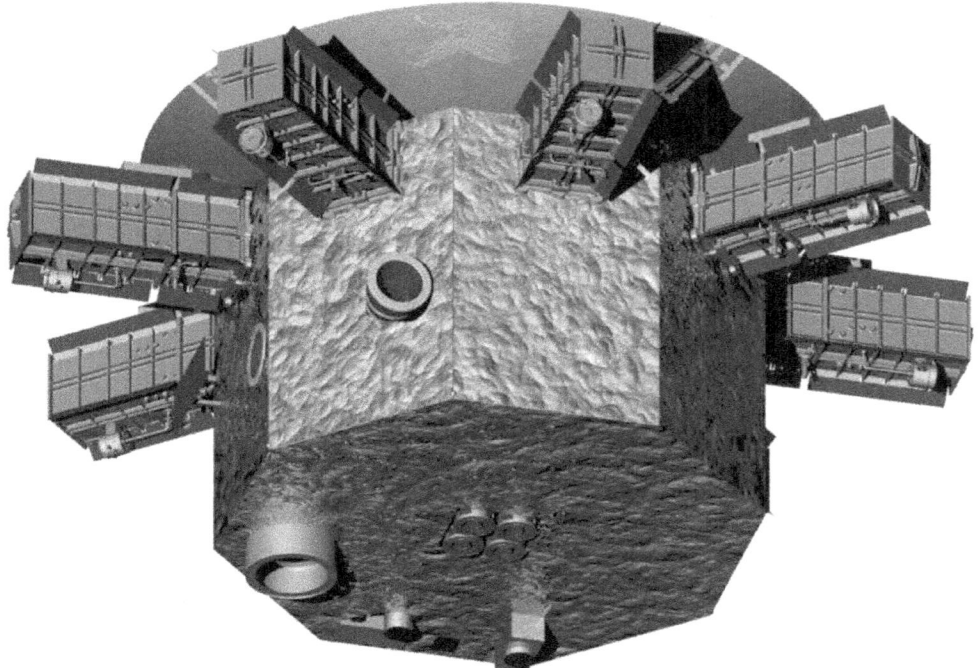

Figure B.2.—REP stage rendered view, showing EP thrusters on bottom face.

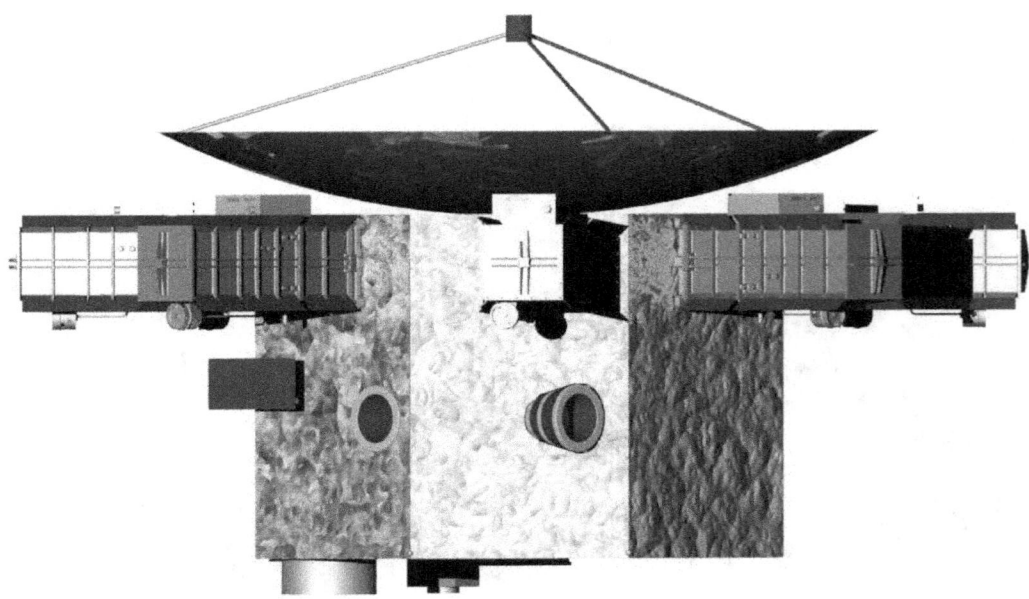

Figure B.3.—REP stage rendered view, face on showing science instruments side.

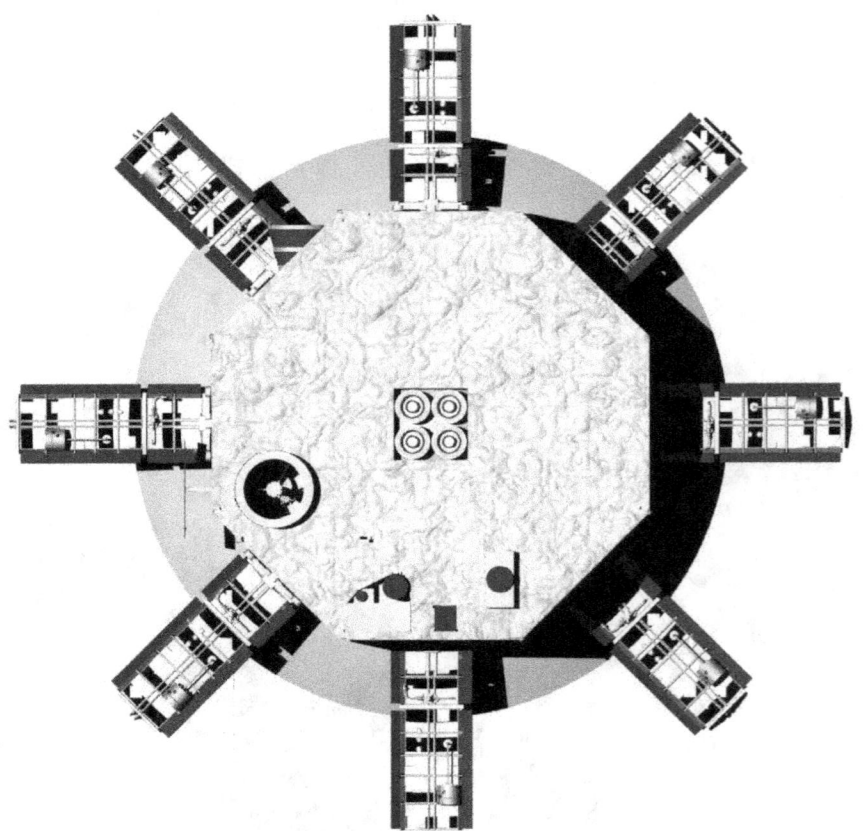

Figure B.4.—REP stage rendered view, showing EP thrusters on bottom face.

Appendix C.—Study Participants

Radioisotope Electric Propulsion (REP) Design Session			
Subsystem	**Name**	**Center**	**Email**
In-Space Program	Len Dudzinski	GRC	Leonard.A.Dudzinski@nasa.gov
In-Space Program	Scott Benson	GRC	Scott.W.Benson@nasa.gov
Lead	Steve Oleson	GRC	Steven.R.Oleson@nasa.gov
System Integration, MEL and Final Report Documentation	Melissa McGuire	GRC	Melissa.L.Mcguire@nasa.gov
Documentation	Les Balkanyi	GRC	Leslie.R.Balkanyi@nasa.gov
Launch Vehicle Integration	TBD	GRC	TBD
Ground Systems	GRC with APL support		TBD
Mission	John Dankanich	GRC	John.W.Dankanich@nasa.gov
Operations, GN&C	Doug Fiehler	GRC	Douglas.I.Fiehler@nasa.gov
Structures and Mechanisms	John Gyekenyesi	GRC	John.Z.Gyekenyesi@nasa.gov
Propulsion	Tim Sarver-Verhey	GRC	Timothy.R.Verhey@nasa.gov
Propulsion	Jim Gilland	GRC	James.H.Gilland@nasa.gov
Thermal	Tony Colozza	GRC	Anthony.J.Colozza@nasa.gov
Power	Paul Schmitz	GRC	Paul.C.Schmitz@nasa.gov
Command and Data Handling	Jeff Juergens	GRC	Jeffrey.R.Juergens@nasa.gov
Communications	O. Scott Sands	GRC	Obed.S.Scott@nasa.gov
Communications	Bin Nyugen	GRC	Binh.V.Nguyen@nasa.gov
Configuration	Tom Packard	GRC	Thomas.W.Packard@nasa.gov
Communications, Avionics and Software	T.C. Nguyen	GRC	Thanh.C.Nguyen@nasa.gov
Cost	Tom Parkey	GRC	Thomas.J.Parkey@nasa.gov
Risk/Reliability	Anita Tenteris	GRC	Anita.D.Tenteris@nasa.gov
Risk/Reliability	Bill Strack	GRC	bstrack@wowway.com

References

1. NASA Solar System Exploration Office (http://solarsystem.nasa.gov/missions/future4.cfm).
2. ANSI/AIAA R-020A-1999, *Recommended Practice for Mass Properties Control for Satellites, Missiles, and Launch Vehicles.*
3. CD–2007–16, "COMPASS Final Report: Radioisotope Electric Propulsion (REP) Centaur Orbiter New Frontiers Mission," Sep. to Nov. 2007.
4. C.G. Justusa, Aleta Duvalla, Vernon W. Kellerb, Thomas R. Spilkerc and Mary Kae Lockwood, "Connecting atmospheric science and atmospheric models for aerocapture at Titan and the outer planets," Available online 12 February 2005. [http://www.sciencedirect.com/science?_ob=ArticleURL&_udi=B6V6T-4FG4VCY-1&_user=141925&_rdoc=1&_fmt=&_orig=search&_sort=d&view=c&_acct=C000011798&_version=1&_urlVersion=0&_userid=141925&md5=e18ccf2b7ce6cd5887132d2d5f155378].
5. M. Lockwood, "Titan Aerocapture Systems Analysis," AIAA–2003–4799, 39th AIAA/ASME/SAE/ASEE Joint Propulsion Conference and Exhibit, Huntsville, AL, July 20–23, 2003.
6. J. Olejniczak, CA; D. Prabhu and M. Wright, N. Takashima, B. Hollis and K. Sutton, "An Analysis of the Radiative Heating Environment for Aerocapture at Titan," AIAA–2003–4953, 39th AIAA/ASME/SAE/ASEE Joint Propulsion Conference and Exhibit, Huntsville, AL, July 20-23, 2003.
7. Lockwood, Mary Kae (2003). "Titan Aerocapture Systems Analysis," AIAA–2003–4799, 39th AIAA/ASME/SAE/ASEE Joint Propulsion Conference and Exhibit. Huntsville, AL, July 2003.
8. Edelbaum, T.N., 1964, "Optimum Low-Thrust Rendezvous and Station Keeping," AIAA Journal, vol. 2, no. 7, pp. 1196–1201.
9. Edelbaum, T.N., 1965, "Optimum Power-Limited Orbit Transfer in Strong Gravity Fields," AIAA Journal, vol. 3, no. 5, pp. 921–925.
10. http://www.adcole.com/en/
11. CD–2007–08, "COMPASS Final Report: Enceladus Solar Electric Propulsion Stage," Mar. 2007.
12. Kerslake, Thomas W.; Haraburda, Francis M.; Riehl, John P., "Solar Power System Options for the Radiation and Technology Demonstration Spacecraft," AIAA–2000–2807; E–12362; NAS 1.15:210243; NASA/TM—2000-210243, July 2000.